THE
DEPENDENT
GENE

THE
DEPENDENT
GENE
The Fallacy of
"Nature vs. Nurture"

DAVID S. MOORE

A W. H. Freeman Book
Times Books
Henry Holt and Company
New York

Times Books
Henry Holt and Company, LLC
Publishers since 1866
115 West 18th Street
New York, New York 10011

Henry Holt® is a registered trademark
of Henry Holt and Company, LLC.

Library of Congress Cataloging-in-Publication Data
Moore, David S. (David Scott), 1960–
  The dependent gene : the fallacy of "nature vs.
nurture" / David Scott Moore.
     p. cm.
  Includes bibliographical references and index.
  ISBN 0-7167-4024-9
  1. Nature and nurture.   2. Phenotype   3. Genotype
environment interaction   I. Title
QH438.5 .M66 2001
576.5'3—dc21   2001002098

First Edition 2002

Designed by Cambraia Fernandes

Printed in the United States of America
3   5   7   9   10   8   6   4   2

CONTENTS

103884

Part V
IMPLICATIONS

Part I

WHERE WE'RE GOING, WHERE WE'VE BEEN

A stone, however large it may be, cannot be enough to build a tall castle. A man, however great he may be, cannot be a hero by himself. A tall castle can be so tall because there are foundation stones that remain unknown. A man can be such a great hero because there are many heroes that remain unknown.

—*Journalist and historian Tokutomi Sohō (1863–1957)*

"Hello, Dolly"
The Future Is Now

Nearly every day at the beginning of this new millennium, we are encountering news reports of the discovery of the gene "for" some human trait or illness. It is easy to understand the excitement surrounding these reports; just consider some of the headlines heralding our new understanding of the human condition:

> "Gene defect may provide cancer test" (*Los Angeles Times*, September 29, 1995)
> "Parkinson's researchers zero in on gene" (*Los Angeles Times*, November 15, 1996)
> "Body's reaction to smog may be tied to genes" (*Los Angeles Times*, March 9, 1997)
> "Mutant gene thwarts HIV" (*USA Today*, August 9, 1996)

In addition, both the scientific journals and the world's leading newspapers now regularly contain reports about the association between certain genes and complicated behaviors or psychological states:

> "Found: A gene that controls place memory" (*USA Today*, December 27, 1996)
> "Is there a gene behind suicide?" (*Los Angeles Times*, August 22, 1996)
> "Variant gene tied to a love of new thrills" (*New York Times*, January 2, 1996)
> "People long on neuroticism appear to be short on a gene" (*New York Times*, November 29, 1996)
> "The 'gay gene' and the politics surrounding it" (*Los Angeles Times*, October 2, 1996)

In fact, there have even been reports of the discovery of a human gene that causes aging!

There is no reason to deny the amazing advances made in the field of molecular biology, the branch of biology devoted to understanding that preeminent biological molecule, DNA. These advances may be among the most important achievements ever in the study of biology, and there is no telling what suffering might be allayed by further study of how our genes contribute to our traits. Unfortunately, these astonishing advances have been presented to the public in a way that has perpetuated the mistaken idea that some of our traits are caused exclusively (or primarily) by our genes. In fact, as we will see, all of our traits—bar none—emerge from the mutually *dependent* activity of *both* genetic and environmental factors. Widespread ignorance of this fact comes with serious consequences.

The origins of our biological and psychological characteristics have been the subject of an enormous amount of scrutiny since Aristotle's time, around 2,300 years ago. And no wonder; in communities around the world, children have perennially been hounded by relatives making exclamations such as, "Oh, you have your mother's eyes!" or "How can you come from *this* family and be so patient?" Ultimately, most of us have an intuitive sense that the key to understanding ourselves is hidden in the origins of our traits. The fact that the processes that give rise to these traits are so inscrutable as to seem miraculous renders them fascinating. But even if these processes were easily discernible, the ways in which our traits develop would still be of the greatest possible importance to us.

The importance of understanding trait development goes well beyond existential questions about how we come to be how we are. Today, as biologists finish mapping great stretches of the human genome, we stand on the threshold of an unpredictable future, one that seems to hold out the possibility of discovering cures for so-called "genetic" diseases, of choosing the traits of our offspring, of identifying violent threats to society before they do any damage, of identifying those individuals who would or would not benefit from various types of public assistance, and even of resurrecting recently deceased loved ones through the stupefying technology of cloning. The technological advances associated with genetic engineering even seem to hold out the possibility of breaking free of natural selection and taking control of our own evolution. Understanding the science behind these technological revolutions is imperative if we, as a society, are to make good decisions about how to best step into the future.

On one level, traditional debates about the relative contributions of genetic and environmental factors to trait development seem strictly academic. After all, most of us already provide our children with developmental environments that are—as far as we know—the best we can offer, so discovering that children are profoundly affected by their environments will not necessarily alter how we parent. Similarly, since we currently have no control over our genes, studying their impacts on our children's traits might seem pointless. But a little further reflection reveals that our positions in this so-called "nature-nurture" debate have all sorts of practical ramifications, whether or not we recognize them.

For instance, it seems that in every election, my community polls its members to see if they are willing to support a school bond issue and thereby commit additional tax dollars to improving our public schools. And while they might not realize it, citizens' opinions about the roles played by environmental factors in intellectual development almost certainly influence how they vote. Some people might reason as follows: If intelligence is largely genetic, perhaps we are wasting our time and money by pouring public funds into our schools. As scientists like Sandra Scarr have surmised, it could be that a school environment that meets certain minimum requirements is "good enough" to allow every student to reach his or her genetically preordained intellectual potential. If we believe this—consciously or not—it will certainly affect our behavior at the polls and will have noticeable effects in our children's classrooms.

Or consider the position of a juror involved in the penalty hearing of Michael Ross, a serial killer who kidnapped, raped, and strangled four teenage girls in eastern Connecticut during the 1980s. In 1987, a jury convicted Ross of these crimes, and he was sentenced to death for his actions. But, after listening to an appeal of his case, the Connecticut Supreme Court overturned his death sentence in 1994, when they determined that prosecutors in the earlier trial had withheld a letter from a psychiatrist that might have helped with Ross's defense. In a subsequent new penalty hearing in 1999, Dr. James Merikangas testified for the defense that scans of Ross's brain had revealed abnormalities, including a smaller-than-normal cerebellum and larger-than-normal ventricles. When told that such abnormalities are characteristic of people who cannot control themselves, what is a juror to conclude? Does Ross's biological condition somehow make him a candidate not for execution but for incarceration in a mental hospital? While many jurors might not even be aware of it, their beliefs about the *origin* of Ross's brain abnormality are liable to significantly impact how they evaluate his culpability. Presumably, jurors would feel one way if the brain abnormalities were the result of a degenerative brain disease, and an entirely different way if those same brain abnormalities were caused by Ross's violent behavior itself.*

The importance of these issues goes beyond the classroom and the courtroom and directly impacts how we run our own lives. For example, some of us have parents who have traits that we consider to be undesirable, such as alcoholism or depression. Given the tendency of such traits to run in families, what should those of us in this position do to avoid succumbing to alcoholism or depression ourselves? Once again, a person's understanding of the *origin* of her traits importantly impacts how she responds

---

*I have referenced all quotations and statements of fact in the Notes section at the back of this book. In the case of notes containing supplementary information that readers might like to peruse, I have indicated the existence of the note with an asterisk in the text.

to having a particular family history. Knowing, as I do, that the heart disease that runs in my family develops, in part, as a consequence of consuming high-fat foods on a regular basis, I have drastically reduced my consumption of ice cream in recent years. But if I thought that obstructed coronary arteries were in my future regardless of my behavior, you can bet that I'd be eating ice cream as if there were no tomorrow! Even for those who never think about trait origins, though, unconsidered assumptions are likely to affect behavior; these effects influence both the shapes of our lives and the lives of those around us.

In fact, how we understand the origins of our characteristics can be a matter of grave importance, having life or death consequences. Alice James, for instance—whose older brothers were the novelist Henry James and the psychologist-philosopher William James—led a short life in the nineteenth century as a sickly, neurotic, and basically unfulfilled sad woman. My mentor Jerome Kagan has recently speculated on the significance for her death of her belief that her temperament was inherited and immutable. He writes:

> The journals of the writer John Cheever, who died in the second half of this [the twentieth] century, and the biography of William James' sister Alice James who died 100 years earlier, imply that both writers inherited a very similar, if not identical, diathesis [i.e., a propensity] that favored a chronically dysphoric, melancholic mood. But Cheever, whose premises about human nature were formed when Freudian theories were ascendant, assumed that his melancholy was due to childhood experiences and tried, with the help of drugs and psychotherapy, to overcome the conflicts he imagined his family had created. By contrast, Alice James believed, with most of her contemporaries, that she had inherited her dour mood. And she concluded . . . that because she could not change her heredity she wished to die.

While Alice James never followed through on her suicidal thoughts, neither did she pursue all of the treatment options that were available to her when it was discovered that she had developed breast cancer; she preferred instead to welcome death as a release from her suffering. Our intuitions about the immutability of our traits cannot help but affect our outlooks, our moods, and our decisions.

THE BRAVE NEW WORLD

In the spring of 1997, the front page of the *New York Times* heralded an unparalleled achievement in the field of biology: a team of Scottish scientists under the direction of Dr. Ian Wilmut had produced the first-ever clone of an adult mammal, a ewe (or adult female sheep) named Dolly.

While clones of *embryonic* mice and sheep had been produced in the 1980s and in 1996, respectively, clones of *adult* animals had proven harder to produce. In the 1950s, tadpoles cloned from adult frogs had managed to live short lives, but they consistently failed to survive to maturity. Thus, prior to the week of February 23, 1997, many biologists had presumed that it would not be possible to produce a healthy adult clone from a complex adult animal.

The events that led to Dolly's conception are truly remarkable. In March 1996, Wilmut's team in Edinburgh scraped a tiny bit of tissue from a ewe's udder (the gland responsible for producing milk). This bit of tissue, like all tissue, was composed of numerous cells—the structural units that make up all living things. These cells were first grown in a laboratory dish, but then, to get the cells' chromosomes into the required state for cloning, the cells were starved of nutrients (note that this is an environmental manipulation, not a directly genetic one). Next, the scientists took some unfertilized egg cells from a different ewe and removed the chromosomes from each one. Finally, they used an electric current to fuse each chromosome-deprived egg cell with one of the intact udder cells taken from the first ewe. Each fusion produced a single new cell—equivalent to a fertilized egg, or *zygote*—which now contained the complete set of chromosomes originally present in the first ewe's cells, and none of the chromosomes originally present in the second ewe's cells. This new cell was then implanted into the uterus of a third, surrogate mother ewe, where it was able to grow just as a normal sheep embryo grows in its mother's uterus. Five months later, after a wondrous series of events *that occur before the birth of all normal mammals and that were completely out of the control of the scientists working in Wilmut's lab*, Dolly was born.

So, popular opinion and lay press notwithstanding, Dolly did not pop out of Wilmut's lab like a copy of this page would pop out of a Xerox machine. No, the story of Dolly's life is significantly more interesting than that, and it belies the commonly held notion that Dolly is a *copy* of anything at all. "Copy" implies that Dolly is an exact replica of the ewe that donated her chromosomes; use of this word in this context reflects the idea that Dolly's characteristics were determined by those chromosomes. But in fact, clones are typically easy to distinguish from the animals that donated their chromosomes, because the characteristics of an animal—be it a clone or a naturally produced animal—are actually determined by the *interaction* of the animal's chromosomes with the unique environment in which the animal develops. Remember, what Wilmut and colleagues produced was *not* a full-grown cloned sheep; rather, they produced a zygote, which when placed in the proper *environment*, developed into Dolly, a full-grown cloned sheep. The developmental processes by which a single cell grows up to be a complex animal remain largely unexplained, but these processes were arguably as important in the production of Dolly as were the actions of the scientists in Edinburgh. I will return to Dolly's tale at the end of the book; for now, it is enough to understand that despite recent technological advances that have allowed for gene discovery, genetic engineering, and cloning, genetic activity still *always* depends on the environmental conditions in which development occurs.

It is rare these days to find someone who believes that complex human psychological traits can be *completely* determined by genetic factors; nevertheless, most people remain unaware that genes alone do not determine the final form of *any* of our traits, even biological traits like eye color. Furthermore, many people think that our genes impose a restricted *range of potential* on us and that the unique characteristics we wind up with *within this range* reflect the impact of our developmental environments. Thus, these people typically believe, for example, that genes can effectively set a person's IQ somewhere between, say, 75 and 120, and that the specific IQ ultimately achieved within this range reflects the environment in which the person develops. But genes do not actually work like that at all. Finally, even when people recognize that it is an error to ask if a trait is caused by genetic *or* by environmental factors (understanding that both play important roles), they still often believe it is possible to ascertain the *relative importance* of each type of factor to the trait's final appearance. Thus, it is commonly thought that some traits can be *more* influenced by genes than can others. But, in fact, it is not possible to parcel causation of traits in this way.

It turns out that knowledge about the true causes of our traits can be acquired in one way only: from studying how traits *develop*. Fortunately, because scientists have been studying biological development for the past 100 years, they now understand some things about trait origins. And their research has revealed at least one incontrovertible fact: *all traits—from "biological" traits like hair color and height to complex "psychological" traits like intelligence—are caused by dependent interactions of genes* and *environments*. Thus, developmentalists openly reject the idea that genes *or* environments can create traits by themselves. Moreover, we maintain that it is theoretically impossible to determine *how much* of a trait is caused by genes and how much is caused by environmental factors. Consequently, the idea that some traits are more—or less—genetically determined than others makes about as much sense to us as the idea that there are monkeys living inside of our heads.

These views—the central theses of this book—have been known by a variety of names, including interactionism, constructionism, dynamic developmentalism, and the probabilistic epigenesis approach, to name just a few; I will refer to them collectively as *the developmental systems perspective*. This view stands in sharp contrast to the views of genetic determinism, which holds that some traits can be caused by genes alone, or at least, more by genes than by the environment. Thus, the developmental systems perspective is radically different from that held by scientists unaware of the facts of development and by almost all of the nonscientists I know. Taken together, the understandings painstakingly gathered from innumerable empirical studies of development have led to a slow revolution in theoretical biology that is likely to end the nature-nurture debate once and for all.

Looking Forward, Looking Back

Few of us think much about development until we near adulthood. We wake up one morning, look at ourselves, and ask, as David Byrne and his band Talking Heads once did, "How did I get here?" We look back over our childhoods trying to understand how we came to be how we are in the present. Thus, we tend to consider development using a strikingly backward-looking approach. I suspect that this perspective influences how people intuitively think about development. We see things very differently when we try to look at development as developmentalists do: in a forward-looking way, seeing the present state as the stage from which we develop.

When we ask ourselves "How did I get here?" we might first think about our experiences, since we are so familiar with the impacts of those experiences. We might next consider how much we look and act like our parents—not only the behaviors they taught us, but the little things that almost seem out of our control, like our gestures and mannerisms. It seems likely that these twin trains of thought about our parents and our experiences lead us to think about development in terms of the independent effects of genes and environments. It is probably no accident that scientists considering this question in the decades before the discovery of chromosomes thought this way, too; after all, with very little data to go by, their thinking was not far removed from the way most of us still think about development.

A different perspective, however, typically yields new insights. When we concentrate in a forward-looking manner on the *causes* of development in maturing organisms, we see a very different story. An incredibly complex sequence of events takes place in the development of every individual, changes that can be seen from this vantage point as being driven by neither genetic nor environmental factors alone but by interactions between the two. When the starting point of one's inquiry is the single fertilized cell that will become first an embryo, then a fetus, then an infant, and, ultimately, an adult, the fact that development is caused by gene-environment interactions becomes plain.

I first encountered the developmental systems perspective in 1990, during a conversation with my colleague-friends Robin P. Cooper and Robert Lickliter,* both professors at Virginia Tech. Like me, these researchers study human infants (Lickliter studies birds as well); we are all fascinated by the origin and development of traits in immature animals. As a result, we were all attending a meeting of the International Society for Infant Studies in Montreal, Canada, when they first described their perspective to me. My initial reaction to their news that genes alone do not cause traits was somewhat subdued. (Years later, I realized that this reaction reflected my failure to grasp their point!) You see, in 1990, like legions of my fellow psychologists who call themselves "interactionists," I thought it was *obvious* that nature and nurture interact to produce traits; this seemed to be the reasonably moderate position lying midway between the more extreme positions held by genetic and environmental determinists, respectively. But in the past decade, I have grown dissatisfied with the facile sort of "interactionism" I used to accept, an interactionism that is built on truisms and that crumbles when examined (as we will see momentarily). In fact, I have come to believe with Lickliter and Cooper that thinking about these issues in the superficial way that leads *most* people to be "interactionists" ultimately leads away from a rich understanding of how people get to be how they are. To produce such an understanding, we need to adopt a more complex and genuine interactionism—a developmental systems perspective—born of the detailed study of *how* traits emerge from gene-environment interactions.

While avoidance of extreme positions leads many people to endorse "interactionism," a little probing often reveals deep-seated misunderstandings about *how* genes and environments interact, a misunderstanding that leaves open the possibility that some traits *are* genetically determined. Such was my misunderstanding in 1990. And as I now do when discussing these issues with my students, Cooper and Lickliter revealed my misunderstanding by querying me about traits that I used to think *were* produced mostly by genes, my self-proclaimed "interactionism" notwithstanding. For example, while many of my students claim to be "interactionists," they typically still believe, as I did, that traits like eye color are determined primarily by genetic factors. It seems that once we leave behind the complex world of human psychological traits, genuine interactionism is often discarded, revealing the unexamined nature of a more facile interactionism. As a result of my journey to understand how genes and environments interact during real-life development, I am now convinced that *all* traits reflect the necessary contributions of both genetic and environmental factors.

To obtain a general impression of the developmental systems perspective, consider the central ideas that have driven the nature-nurture debate for the past 130 years. As traditionally conceived, "nature" is taken to refer to "human nature" and is imagined to reside within our bodies, to be that with which we are born, that which exists independently of our experiences. In contrast, "nurture" is taken to arise outside of our bodies; usually, "nurture" is understood to comprise our experiences. But a little further reflection shows these conceptualizations to be problematic. While we are certainly nurtured by elements of our environment, our bodies (and minds!) are also shot through with that nurture. Every breath we take, every meal we eat, every scene we see is assimilated into the very structures and functions of our bodies, literally becoming us. And these processes of incorporating our environments into ourselves begin at conception and continue throughout our lifetimes; they operate as our bodies and minds are literally built from "scratch" through the mutual actions of those natural elements that are within and around us. When we first see the light of day as newborn babies, it is already impossible to identify any part of us that does not reflect the environment in which we developed from conception.

The same story holds for nature: we are certainly natural, but we are also completely embedded in nature. The sun, the trees, and the clouds that constitute our environment are no less a part of nature than are our genes, our eyes, and our sorrow. Even nurture is natural. This can hardly be denied: *We are shot through with nurture even as we are embedded in our nurturing environments, and we are embedded in nature even as we are shot through with it.* The developmental systems perspective finds the traditional nature/nurture dichotomy to be false and misleading, a dichotomy that needs to be eliminated before any real understanding of ourselves can be achieved. The rest of this book looks at the details of real development and shows these details to be perfectly consonant with the above sketch.

Unfortunately, many of the studies underpinning the developmental systems perspective are loaded with jargon born of recent advances in biology; many of them would be incomprehensible to those without training in this field. Still, given the far-reaching implications of the insights culled from these studies, it is important that the public be given access to them. My purpose is to present these studies—and the insights they have generated—in a way that will make them intelligible to all interested parties. My hope is that my training as a developmental psychologist has left me able to interpret these ideas for nonbiologists. After all, as a nonbiologist, I know what it's like to think about biology as nonprofessionals do.

Even so, fair warning is warranted here: while I will use as little jargon as possible and always define and explain new terms that I think must be introduced, grasping these new ideas will be challenging. Surprisingly, though, the biological content of this book is not particularly difficult. Instead, the challenge arises from the fact that the developmental systems perspective requires the rejection of ingrained ways of thinking that we all have been exposed to, continuously, for many decades. Nondevelopmentalists with whom I have spoken about this perspective have initially had a rather healthy skepticism for an approach that insists that eye color is no more determined by genes than is personality. And no wonder; we have all been steeped in grade-school understandings of genetics for all of our lives, and so we are quite confident that a blue-eyed man's blue eyes were *caused* strictly by recessive genes that he inherited from his parents. The challenge of this book will involve relinquishing skepticism that is, in this case, unwarranted. Training oneself to think in a novel way is never easy, but I trust that my readers will enjoy the challenge. The developmental systems perspective reveals the beauty inherent in the development of living things. It instills optimism in anyone who has struggled successfully to understand it and so is—at its best—truly liberating. Even at its worst, it is fascinating.

A PRIMARY STRATEGY, A FUNDAMENTAL ASSUMPTION, AND AN OVERRIDING BELIEF

As with most books, there are strategies, assumptions, and beliefs underlying this one; to my mind, some of these warrant disclosure—or possibly, justification—from the outset. So, before proceeding, I would like to address these three issues in turn, starting with one of my strategies. Throughout the book, I will often illustrate gene-environment interactions by discussing how relatively *simple* animals develop *biological* traits. My use of this strategy flows from two observations, the first of which gives rise to my focus on biological traits, and the second of which gives rise to my focus on simple animals.

Most people seem to me to be more open to the idea that gene-environment interactions produce psychological traits than to the idea that such interactions produce biological traits. While biological traits such as eye color often appear to be impervious to environmental influences, it is usually apparent that biological and environmental factors both contribute *something* to the development of psychological traits. For example, you need certain experiences to learn Spanish, but if you didn't have a brain, you wouldn't learn any language at all. If interactionist explanations are easier to accept for psychological traits than for biological traits, then an account that demonstrates that *biological* traits are caused by gene-environment interactions will probably leave a reader as convinced as ever that psychological traits, too, are produced by interactions, and not by genes alone. For this reason, I will often focus my discussions on the development of biological traits.

Second, the traits of simple animals sometimes appear to be less affected by their developmental environments than are the traits of more complex animals. For example, a particular type of nematode (a kind of worm) is about 40 times simpler than a human being—it has 40 times fewer identified genes—and each male nematode has exactly 1,031 cells that constitute his body. This invariant number of cells produces the impression that all normal worms of this type develop in exactly the same way, independently of environmental factors. In contrast, more complicated animals like fruit flies have bodies that develop in ways that are clearly influenced by nongenetic factors, and complex animals like people develop in ways that exhibit little of the environmental independence that *appears* to characterize nematode development. Hence, scientists sometimes conclude that the developmental environment affects the traits of simple animals less than it does the traits of complex animals. Given this conclusion, a demonstration that *simple* animals' traits develop via gene-environment interactions should leave readers convinced that the development of complex animals' traits, too, depends at least as much on such interactions.

Because of this state of affairs, I will often illustrate my points by describing the development of biological traits in relatively simple animals like sea urchins, fruit flies, and worms. When encountering these illustrations, keep in mind what they mean for human traits, of both the biological and psychological varieties. If the biological characteristics of fruit flies are not caused primarily by genes, the biological and psychological traits of people are probably not caused primarily by genes either.

But never forget: *the developmental systems perspective does* not *hold that environmental factors are* more *important than genetic factors in the production of any trait.* Because these days we are more likely to be misled by the wrong-headed notion that genes alone can produce traits than we are by the equally wrong-headed notion that the environment alone can produce traits, I will focus on the importance of environmental factors in trait development. Nonetheless, the developmental systems perspective maintains that genetic and environmental influences are always equally important in producing *all* of our traits.

This tenet leads to a fundamental assumption that I will make throughout this book, namely that we can ignore the distinction traditionally drawn between biological traits like height or facial structure and psychological traits like temperament, intelligence, or personality. There are several reasons for assuming this distinction can be ignored. First, most contemporary philosophers and psychologists agree that the psychological mind is a product of the biological brain (this agreement explains the recent tendency for research psychologists to be drawn to biological explanations of psychological phenomena). People currently working in these fields, by and large, believe that behavior can be understood to be an aspect of biology. Second, there is an ever-growing body of evidence suggesting that psychological states directly impact our biology (at the level of the body's organs, as well as at the level of the body's cells and their DNA). Thus, biology and psychology mutually influence one another. Third, and perhaps most important, the ideas I will present apply equally well to the origin of both biological and psychological traits, reinforcing my decision to ignore the distinction. Since the developmental systems perspective holds that genetic and environmental factors are equally important in the production of all traits, it should be clear why, for the purposes of this book, it makes sense to refrain from drawing firm distinctions between psychological and biological traits.

Finally, the way I think about trait causation makes sense only in the context of a particular belief that I think most developmental systems theorists share, namely that the goal of science should be *comprehension that allows for effective intervention in nature's doings.* We are obliged to understand natural phenomena at this level because of the fact that we can never *know*, for sure, if our theories about nature are accurate. Unfortunately, an adequate discussion of this philosophical point is beyond the scope of this book, but inasmuch as William James—the father of American psychology—carefully considered this issue in his book *Pragmatism* a century ago, a quotation from this work is in order. In 1907, James wrote that scientific theories appear

> less as a solution . . . than as a program for more work, and more
> particularly as an indication of the ways in which existing realities may be
> *changed. Theories thus become instruments, not answers to enigmas* [emphasis in
> original]. . . . We don't lie back upon them, we move forward, and, on
> occasion, make nature over again by their aid.

As will become apparent, this emphasis on intervention has important implications, both for public policy and for how we think about the causes of our traits. More specifically, it implies that we should strive for an understanding that allows us to *affect* trait development (presumably in ways that are compatible with the democratically determined values of our societies). And because such an understanding requires us to ask *how* genes and environments interact during development, we are encouraged to pursue answers to *this* question, not questions about *which* factors—genetic or environmental—are of most importance in determining our characteristics.

While it would clearly be useful to have an understanding of the ways in which our traits really develop, there are still an enormous number of riddles left to be solved in this puzzle. Nonetheless, at this point in history, some things about the origins of our traits are understood. And given technological advances in the life sciences that will—*in our lifetimes*—allow us to request specific genes for our offspring, to request that a Parkinson's-afflicted parent receive brain transplants of tissue derived from human embryos, or to request that information about our chromosomes be deleted from an insurance company's database, it is more pressing than ever to understand the role of genetic factors in trait development. But since today's understanding has been built, brick by brick, on older understandings, it will be useful to begin exploring these questions by considering some of the more important advances in the history of ideas about trait origins.

1

FROM ARISTOTLE'S WONDER TO A FORK IN THE ROAD

The Wrenching of Genetics from Development

On a recent visit to Cambridge, England, I walked into the town's oldest standing pub, The Pickerel, and encountered the following words painted on the wall:

> 'tis not to be doubted but [members of the species of fish known as Pike] are bred . . . of a weed called pickerel weed. . . . This weed and other glutinous matter, with the help of the sun's heat, in some particular months, and some ponds adapted for it by nature, do become Pikes. But doubtless, divers Pikes are bred after this manner, or are brought into some ponds some such other ways as is past man's finding out.

This passage comes from a seventeenth-century book by Sir Izaak Walton entitled *The Compleat Angler;* clearly, Europeans at that time were still perfectly baffled by the processes that give rise to new generations of living things (although they were already quite adept at constructing agreeable drinking locales). A modicum of clarity in this domain required at least 200 additional years of incisive thought and research.

It can be no surprise that the arrival of offspring has bewildered humankind since antiquity: reproduction and development are processes that are amazing and seemingly unfathomable. Against this backdrop, the understandings of conception and development shared by contemporary scientists are all the more striking. The intellectual journey that has led to these understandings is quite a tale in itself.

As Aristotle initially outlined the philosophical problem of biological development in *Generation of Animals,* an organism's form could develop in two conceivable ways: either a minute version of the individual could already be present in the fertilized egg (in which case development amounts merely to an increase in size of an already formed mini-organism), or the

16

organism's form could develop—one feature at a time emerging from initially featureless tissue—as a result of interactions between the fertilized egg and its environment. By the eighteenth century (when biologists were maximally embroiled in this debate), the first of these hypotheses was known as *preformation* and the latter as *epigenesis*. Aristotle was quite clear that he believed epigenetic processes were responsible for final biological forms, but his position continued to lack hard scientific evidence for millennia. In the meantime, before the controversy was resolved, the preformationists split into two opposing camps: the spermists, who believed that each sperm carried a miniature organism called a homunculus, and the ovists, who believed that the homunculus originated in the egg.

In retrospect, it seems almost unbelievable that many credible early-nineteenth-century scientists were preformationists. After all, as Gilbert Gottlieb has noted, the Russian biologist Caspar Friedrich Wolff had given us "the empirical solution of the preformation-epigenesis controversy" as early as 1759, providing "the necessary direct evidence for the epigenetic ... [nature] of individual development" by examining the development of actual chick embryos. Failing to find anything that looked remotely like a miniature preformed chick, Wolff had concluded that development occurs epigenetically. Furthermore, as most modern textbooks on the subject are quick to point out, preformationism on its face entails a curious problem: if each egg contains a complete homunculus, mustn't there be another complete homunculus in each of *that* homunculus's eggs?* And mustn't *those* homunculi have eggs each containing their own even smaller homunculi? In fact, mustn't Eve's body have contained all of the people ever to inhabit the planet, to the end of human history?

The questions raised by this problem and by direct embryological observations are typically taken to be the Achilles' heel of preformationism. But in a marvelous book entitled *Ontogeny and Phylogeny*, Stephen J. Gould goes to some length to clarify why this position remained popular into the nineteenth century. First of all, it was not yet believed that the world's history would take more than a few thousand years to unfold; it was much easier for nineteenth-century scientists to imagine each of Eve's eggs containing the limited number of generations needed to fill this relatively short time than it is for us to imagine her eggs containing the enormous number of generations now known to have already inhabited the Earth. Secondly, scientists of the time knew nothing, yet, of the lower limits to the size of cells. At the time, protozoa invisible to the naked eye had just been discovered with the help of that fabulous new invention, the microscope. Given the possibility of inventing ever-better tools for exploring microworlds, it was not irrational for scientists of the time to think that future technologies would reveal ever-smaller creatures; in such a context, the idea of homunculi within homunculi within homunculi would seem significantly less ridiculous.

Preformationists also thought future technologies might invalidate the direct embryological observations of the day; perhaps, they argued, Wolff *saw* the embryonic chick's organs developing in sequence because his technology wasn't sophisticated enough to allow him to see an organism as small as a minute, preformed chick might be. Given this state of affairs, we can see that it was not entirely unreasonable for some scientists to cling to preformationist tenets until such tenets were *proven* to be erroneous.

Because of the limits of eighteenth-century microscope technology, no one had ever actually laid eyes on a mammalian egg prior to 1827. In that year, Karl Ernst von Baer reported that he had discovered that female mammals, too, actually do have eggs. One year later, von Baer published his meticulous observations of developing chick embryos. Suddenly, a consensus began to build in the scientific community that preformationism had to be wrong. In his masterpiece *On the Development of Animals*—widely considered to be one of embryology's most important works of all time—von Baer wrote:

> The more homogeneous the entire mass of the body, the lower the stage of development. We have reached a higher stage if nerve and muscle, blood, and cell-material are sharply differentiated. The more different they are, the more developed the animal.

These empirical observations left most biologists convinced that there are, in fact, no preformed homunculi residing in either eggs or sperm; otherwise, undeveloped embryos would not be as homogeneous as von Baer had observed them to be. Development, it was concluded, must be a fundamentally epigenetic process.

MEET THE NEW BOSS, SAME AS THE OLD BOSS

One might expect our story to end here, as the preformationists are vanquished by the painstaking descriptive work of an exemplary embryologist. But old ideas die hard, and very, very old ideas are even more obstinate. As the nineteenth century progressed, several scientists advanced novel theories of heredity that were still decidedly preformation*istic*, despite the fact that they explained trait origins in ways that avoided positing the existence of actual homunculi. The most important of these theories began to be formulated in 1883 by the German biologist August Weismann, who called his hypothesis "germ plasm theory" ("germ" in the sense of "initial stage," as opposed to in the sense of "virus" or "microorganism").

According to Gottlieb:

> August Weismann's germ plasm theory of heredity held that . . . the fertilized egg or germ contained in it all the necessary "information" for the construction or assembly of the organism. More specifically, Weismann . . . and many other scientists of the late nineteenth century believed the germ to be a highly complicated structure whose various parts ("determinants") corresponded to [and gave rise to] all the organs of the future organism. . . . The basic idea was that the germ plasm consisted of a very large number of [determinants], each corresponding to a unit of the body.

In addition, Weismann proposed that "not all determinants . . . enter every cell of the embryo; instead . . . the [determinants] were hypothesized to divide in such a way that different . . . determinants entered different cells." In this way, some cells would receive information that would allow them to develop into one type of organ (for instance, a heart), while other cells would receive information that would allow them to develop into a different type of organ (for instance, an eye). Thus, while the traditional preformationist notion of a homunculus had been abandoned, a "neo-preformationist" idea that the sperm or egg was prepacked with all the necessary *information* to build a complete body—independent of any but the most trivial interaction with the environment—was still alive and well. Weismann acknowledged as much in his 1894 treatise *The Effect of External Influences upon Development*; here, he wrote that he believed "in a preformative arrangement of the germ-substance."

Weismann's germ plasm theory, which bears a striking resemblance to the modern notion that traits can be caused by genes, is defensible until one actually does experimental studies of trait development in animals. And by 1891, a German biologist named Hans Driesch—one of my personal heroes—had completed a series of such studies in an effort to evaluate the validity of Weismann's neo-preformationistic ideas. But there is a very good reason that experimental studies of embryos were not conducted until the end of the nineteenth century, and it's a story worth telling.

UNSCIENTIFIC MYSTICISM

In the hundred years following the publication of Wolff's observations of chick embryos, several scientists—von Baer among them—did report studies on the embryological development of many species. But, as is often the case in new sciences, these initial papers were strictly descriptive; no one was yet asking what *caused* development. Then, with Charles Darwin's 1859 publication of his theory of evolution in *On the Origin of Species*, the theoretical orientation shared by most biologists for centuries was modified in such a way that the cause of development was *assumed* to be evolution, without anyone having bothered to actually study development experimentally.

In the centuries before 1859, scientists ascribed great theoretical importance to the observation that the stages organisms pass through during development seem to parallel the ascending sequence of *types* of animals—from reptiles, through eels, to fish, through flying fish, to birds, through bats, to mammals. Gould notes that this hypothetical "sequence of increasing perfection ... was entirely static. It had been created (by God, it was assumed) all at once, and its constituent links could not be transformed one into the other."

Then, in 1859, along comes Darwin, arguing quite convincingly that species can *change* over long periods of time. Suddenly, the parallels between the developmental stages in the lives of individual organisms and the changes in species across generations took on new significance. Just seven short years after the publication of Darwin's theory, Ernst Haeckel was busy proselytizing the idea that "ontogeny [the development of an organism over its lifetime] is the short and rapid recapitulation of phylogeny [the evolution of a species across generations]. . . . During its own rapid development . . . an individual repeats the most important changes in form evolved by its ancestors during their long and slow paleontological development." By 1874, Haeckel was proclaiming that "Phylogenesis [evolution] is the mechanical cause of ontogenesis [individual development]." As a result, there was no need to do experiments on embryos to discover the causes of their development; it was assumed that development was caused by evolution.*

These days, no one believes that Haeckel was correct. But a bit more consideration of his idea that evolution causes development will help illuminate some of the problems associated with trying to determine "causation." And it was a desire to explain development with reference only to particular types of causation that ultimately led to the *experimental* study of embryos, eventually casting doubt on the value of any preformationistic theory, newfangled or otherwise.

Most of us think of human development as having an end state (usually, the state the theorizer is in!). Typically, we don't even question the assumption that the adult state is the *goal state*, the state toward which all normal development proceeds. Explanations of development that refer to the final state toward which development proceeds are called *teleological* explanations (from the Greek word *telos*, meaning "end"). For example, a teleological explanation for the form of an eye would refer to the mature eye's usual *function*; eyes are formed as they are be*cause* the function of mature eyes is to provide the brain with clear visual images of the world, and to perform this function, mature eyes have to be formed as they are (that is, with a small hole—or pupil—in front, and light-sensitive cells— the retina—in back). Teleological explanations are so familiar to most of us that it is sometimes hard to even imagine how one might explain a phenomenon like development *without* reference to a final, mature state that serves a particular function.

Likewise, Haeckel couldn't imagine explaining development nonteleologically. Because he thought that natural selection—Darwin's proposed mechanism of evolution—caused the developmental processes responsible for traits, he assumed that most (if not all) traits have adaptive functions. That is, they *do* something that increases the likelihood that animals with those traits will survive and reproduce. For example, Haeckel accepted that giraffes have long necks because natural selection conferred upon long-necked giraffes an advantage over shorter-necked giraffes in the quest for survival; in particular, long-necked giraffes can reach tree leaves that are higher than those that can be reached by shorter-necked giraffes. In explaining the causes of the development of this (and every other) trait, Haeckel would always refer—teleologically—to the *function* of the mature, adaptive trait, the function toward which development was directed; today's giraffes have long necks be*cause* long necks *function* to help giraffes survive (and ultimately reproduce).

But there are other types of causation. In particular, Aristotle proposed that there are four distinct types of causes, known as formal causes, material causes, efficient causes, and final causes. The formal cause of the characteristics of a chair, for example, is the design of the chair; the chair is shaped as it is be*cause* the blueprint for the chair called for it to be constructed with that shape. The material cause of a chair's characteristics, in contrast, is the wood that the chair is made of; the chair has the characteristics it does, in part, be*cause* it is made of wood. The efficient cause of a chair's characteristics is the behavior of the carpenter who built it; the chair has the characteristics it does be*cause* of the carpenter's specific actions (this is the "normal" kind of causation with which we are all familiar). Lastly, the final cause of the chair's characteristics is the chair's *function*; the chair exists as it does be*cause* the carpenter was desirous of a comfortable seat. As such, final causes—and only final causes—are teleological.

In the 1880s and 1890s, some members of the scientific community began to consider the possibility that biological development could be understood without reference to teleological explanations at all. For instance, in contrast to Haeckel, who would have argued that the development of an eye is best explained with reference to the *function* of mature eyes, these scientists believed that teleological explanations would be unnecessary if the *efficient* causes of development could be revealed. What, they wanted to know, are the antecedent conditions that enable an eye to develop? How, exactly, is an eye actually *constructed* in development? In 1888, one of these scientists—Wilhelm His—wrote:

> The single word "heredity" cannot dispense science from the duty of making every possible inquiry into the mechanism of organic growth. . . . To think that heredity will build organic beings without mechanical means is a piece of unscientific mysticism. . . . By comparison of different organisms . . . we throw light upon their probable genealogical relations, but we give no direct explanation of their growth and formation. A direct explanation can only come from the immediate study of the different phases of individual development.

Clearly, His believed that scientists should perform experimental manipulations on organisms in different phases of development, to determine what sorts of *efficient* causes lead from one phase to the next. And since—using His's words—"every stage of development must be looked at as the physiological consequence of some preceding stage," it makes sense to begin studying the efficient causes of development right at conception. With the advent of this approach to the study of development, teleological explanations of the sort favored by Haeckel fell out of fashion, and a new breed of experimental embryologist was born.

THE BIRTH OF A SCIENCE, THE DEATH OF AN IDEA

The title of "father" of the new field of experimental embryology is traditionally bestowed upon the German biologist Wilhelm Roux, who founded a journal in 1894 to publish the work of experimental embryologists like His. The researchers who followed the path laid out by His and Roux

> experimented by disturbing the normal course of development; they studied embryonic stages to discover their proximate causes in previous conditions and to assess their influence upon following ones. . . . Experimental embryologists relentlessly asserted that their kind of cause [efficient cause] exhausted the legitimate domain of causality. All that had come before them was merely descriptive; they had established the first causal science of embryology. . . . Thus, Wilhelm Roux began the prolegomenon to his new journal—*Archiv für Entwicklungsmechanik der Organismen* ["Archive for the Developmental Mechanics of Organisms"]— with these words: "We may designate as the general goal of developmental mechanics the ascertainment of formative forces or energies."

Thus, Roux, His, and their followers were adamant about explaining development without using Haeckelian teleological explanations.*

Roux was one of August Weismann's contemporaries, and he believed that Weismann's germ plasm theory was probably correct. So, in 1888, in the first truly experimental investigation of embryological development, Roux set about putting that theory to the test.

Given that frog eggs are easier to observe than just-visible-to-the-naked-eye human eggs (since, at 1–2 millimeters in diameter, they are 10–20 times bigger), Roux decided that they were perfect for studying the early development of vertebrates (that is, animals with skulls, brains, and backbones, like frogs and humans). Roux obtained a number of fertilized frog eggs and then watched them until they divided into two cells, thereby completing the first step in the development of all multicellular animals. He then used a hot needle to kill one of the two resulting cells, thinking that if Weismann was correct, the hot needle would destroy the "determinants" that corresponded to half of the body's organs. Next, he let the single undisturbed cell continue to develop (while still attached to the other, now dead, cell). And just as he expected, given his faith in Weismann's theory, this procedure produced a half-embryo; sometimes he wound up with a frog embryo with a tail region but no head region, and other times he wound up with a head region but no tail region. Regardless, the resulting embryo was always abnormal, suggesting—as Weismann's theory predicted—that pre-existing "determinants" *had* resided in the original fertilized cell, that those "determinants" specifying the head and tail regions had wound up in different cells when the original cell had split in half, and that half of the "determinants" had been destroyed by the hot needle when the embryo was in the two-cell stage. Such a demonstration was unprecedented.

Nonetheless, Hans Driesch was unimpressed. While Driesch was a firm believer in the program of the experimental embryologists*—and so applauded Roux's experimental approach—he thought Roux's results and interpretations deserved a bit more scrutiny. Driesch recognized that "Weismann's idea was only a little less crude than the earlier belief in preformation," and, as a result, he found it hard to abide by Roux's claims without first collecting additional experimental data of his own. In the next three years, he conducted an ingenious series of variations on Roux's experiment, publishing the results in 1891.

Driesch thought it was possible that Roux's hot needle technique was too rough for fragile frog embryos; he thought that maybe Roux's outcomes were caused by inadvertent damage to the healthy cell when the other cell was being pierced, as opposed to being caused by the purposeful damage done to the pierced cell. Therefore, Driesch decided to do Roux's experiments on sea urchin eggs, because he knew that these were much less fragile than frog eggs. He took sea urchin embryos that had reached the two-cell stage, and he shook them until the two cells separated from one another. The next morning, Weismann's preformationistic germ plasm theory was relegated to the cosmic dustbin of rejected ideas; from *each* single cell, a complete, normal sea urchin embryo had developed—two for the price of one! Clearly, the original zygote cannot have contained a single set of bodily "determinants" (let alone a fully formed homunculus!) that was divided when the zygote split in half, or Driesch's result would have been impossible. Rather, as we now know, all cells in very young embryos are "equipotential" (to use Driesch's word), meaning that *each cell, given the proper environment*, has the ability to produce a complete, normal organism.

As it happens, Driesch was *not* right about the cause of Roux's experimental results. Roux's healthy cell wasn't damaged when he pierced the other cell in the two-cell embryo, but its subsequent development was abnormal because of the continued presence of the other, now dead, cell. When one of two cells in a normally developing frog embryo is killed by a hot needle and then *removed*, the remaining cell does develop into a normal, complete frog; therefore, Driesch's results apply to vertebrates as well as to simpler organisms like sea urchins. In fact, the results apply to mammals such as mice and humans, too.

In some ways, this story is even more amazing for mammals. When Driesch took a sea urchin embryo that had developed to the four-cell stage and shook it hard enough to separate the four cells from each other, he ultimately wound up with *four* complete sea urchins. This finding strengthened his original conclusions, of course. But if a sea urchin embryo is allowed to develop to the eight-cell stage and then the top four cells are separated from the bottom four cells, two complete embryos are *not* produced; this procedure yields one slightly abnormal embryo, and one hollow ball of cells. Apparently, by the eight-cell stage, something about the top four cells of a sea urchin embryo is different from the bottom four cells. But such is not the case for mammals this early in development. Even at the sixteen-cell stage, the cells that make up young mouse embryos can be rearranged "in numerous combinations and normal development will still occur. Even if several mouse embryos are pushed together so that they fuse . . ., a [single] normal mouse will still develop." And inasmuch as identical human twins are typically produced *not* by disassociation and subsequent development of the two cells in a two-cell embryo, but by division of a single embryo composed of *hundreds* of cells into two masses of cells, each of which goes on to develop into its own complete person, human embryos must be made of cells that retain their equipotentiality for a relatively long period of developmental time. So, the conclusions of Driesch's sea urchin experiments are on firm ground when it comes to people as well as invertebrates: a zygote cannot contain a fully formed homunculus, or even "determinants" for body parts that are split up when the zygote divides. Therefore, preformationism—and all ideas related to it—must be incorrect, and development must be an epigenetic process.

Driesch considered the results of his variation on Roux's experiment to have sounded the death knell for both Weismann's theory and for preformationism in general. Nonetheless, by the time he published the results of his revolutionary experiments, the tenacious seeds of neo-preformationism had already been sown. They were merely lying dormant, awaiting a metaphorical rain that would allow neo-preformationism to burgeon into a genetic determinism that, a century later, still pervades many people's thinking about the origin of traits. This rain arrived shortly, in the form of the rediscovery in 1900 of work published 34 years earlier by a rather unusual monk named Gregor Mendel.

THE FOUNTAINHEAD OF GENETICS

Gregor Mendel became a monk in 1848, after entering a monastery at the age of 21 in what is now the Czech Republic. When his subsequent work as a substitute high school science teacher failed to blossom into a career—he was unable to obtain certification as a regular teacher given his low examination scores in zoology—he went to Vienna to study both science and statistics. Statistics was just emerging at that time as a brand new branch of mathematics.

At the time, the only accepted theory to explain the transmission of traits from generation to generation was the theory of *blending inheritance*. The idea—accepted by everyone, including Charles Darwin—was that the traits of offspring are produced by blending together the traits of their parents, so that a tall man and a tall woman would have tall offspring, a short man and a short woman would have short offspring, and a tall man and a short woman would have offspring of intermediate height. But unlike his contemporaries, Mendel was troubled by this idea (perhaps this explains his poor grades in zoology!). The difficulty with the theory as Mendel saw it was that it was very obviously wrong; while sometimes traits do seem to blend—for instance, mating plants with red flowers and plants with white flowers sometimes *does* produce plants with pink flowers—mating a blue-eyed man with a brown-eyed woman never produces offspring with blue-brown eyes. There's even a problem when blending inheritance does explain the outcome of a first-generation mating. This theory predicts that if plants with pink flowers—produced by mating plants with red flowers and plants with white flowers—are now mated with one another, all the second-generation offspring would have to have pink flowers; after all, a blend of pink and pink is pink. But in fact, such a mating often produces plants with flowers as purely white or as purely red as the original (grand-parent) plant's flowers. So, Mendel thought, the theory of blending inher-itance has to be wrong.

Back in the monastery gardens in 1856, the 34-year-old Mendel began to conduct a series of scientific experiments to address the problems that he thought were inherent in the theory of blending inheritance. In these studies, he crossbred different strains of common garden pea plants and examined the traits inherited by the first-generation offspring. He then mated the first-generation offspring to one another and examined the traits of the second-generation offspring. Finally, he mated the second-generation offspring to one another and examined the traits of the third-generation offspring. In order to comprehend the results of these experi-ments, Mendel felt forced to postulate the existence of "heritable factors" that determine the traits of offspring; he called these factors *Formbildung-elementen*, or form-building elements. His data suggested to him that the *Formbildungelementen* should be regarded as irreducible and material parti-cles that could not blend with one another. Further, he assumed that they existed somewhere within the gametes (that is, sperm and eggs in animals, or ovules and pollen in plants).

This was the crux of a report Mendel read in 1865 to a group of local scientists; we can surmise from their public reactions to his speech—there were none—that they were unimpressed. The following year, Mendel pub-lished a paper reporting his results and his explanations for his results; again, his announcement was met by silence from the larger scientific com-munity. Mendel died a quiet death in scientific obscurity eighteen years later, in 1884. And when Weismann published his germ plasm theory the following year, no one had ever heard of Gregor Mendel.

Then, in 1900, three different scientists in three different countries, each of whom was trying to understand the nature of biological inheri-tance, simultaneously discovered Mendel's heretofore ignored paper and immediately understood that he had solved one of biology's most perplex-ing puzzles 35 years earlier. Almost overnight, and posthumously, Mendel took the scientific world by storm, becoming one of history's most famous scientists—the world's first geneticist.

When Mendel's ideas were rediscovered, many biologists could not resist the conceptual similarity of Weismann's theoretical "determinants" and the "heritable factors" that Mendel had so brilliantly utilized to explain his data; the two concepts just seemed to map perfectly onto one another! The upshot was that most of the scientific community began to think of Mendel's *Formbildungelementen* (which later came to be called genes) in the same way that Weismann thought about his hypothetical "determinants"—as "self-contained packets of inheritance" that each correspond to, and give rise to, specific traits. And in the marriage of these two ideas lay the cornerstone of genetic determinism, a preformationistic idea that was able to somehow survive even Driesch's already published, unequivocal empirical refutation of such ideas.

As noted in one of today's most popular textbooks of child development:

> In the nineteenth and early twentieth centuries, the old theory of
> preformation was considered to have been little more than a mystical
> substitute for science. Currently, however, it is being remembered more
> favorably. . . . [What preexists in the sperm and egg] is not a minuscule,
> preformed human being, as the early preformationists thought. Rather,
> what preexists in the zygote is now understood to be a set of coded
> instructions contained in the genes.

Thus, this neo-preformationism is one in which the *information* needed for the production of a body is completely contained within the sperm and egg, in which individual genes invariably *determine* individual traits (hence the term "genetic determinism") and in which the body's development occurs independently of its environment and experiences.

Genetic determinism dismisses Driesch's conclusions with the hypothesis that the equipotentiality of embryonic cells results from a process wherein genes are *copied*, so that each time a cell divides, the newly created cells each receive a copy of the entire set of genetic "determinants." But such a hypothesis cannot explain all of the embryological data. For one thing, it fails to explain why the cells that Roux did *not* pierce developed into *half* embryos as a result of being attached to a pierced (dead) cell; such a procedure should make no difference if the undisturbed cell contains a full complement of genes and develops independently of its environment. For another, if all newly created cells contain the same genes, how do some cells ultimately develop the characteristics of, say, heart cells while others develop the very different characteristics of, say, brain cells? But the beginning of the twentieth century was not the time to dwell on such problems; the new, exciting, and promising science of genetics had just appeared on the horizon.

Shortly after 1900, extremely quick progress was made in the development of microscopes, allowing scientists to actually examine some of the minute structures contained within cells. Before long, great strides were made in identifying cellular structures that could have served as Mendel's heritable "factors." Extremely large molecules called chromosomes were among the best candidates, because scientists had observed that there was a strong correlation between an animal's sex and characteristics of some of its chromosomes. But in 1910, a scientist named Thomas Hunt Morgan wasn't convinced.

SCHISM!

It is well known among scientists that strong correlations reveal nothing about causation. A strong positive correlation means only that as the measure of one factor (say, height) increases, the measure of the correlated factor (say, weight) also increases (and vice versa, so that short people are not particularly heavy, on average). Whenever two factors are strongly correlated, we know that either (a) the first causes the second, (b) the second causes the first, or (c) some third factor causes both of the other two. Unfortunately, *mere discovery of a correlation never allows us to infer which of these three is correct.* For example, if a high correlation is detected between poverty and the presence of schizophrenia, that does not tell us if poverty causes schizophrenia (possibly via the stresses that characterize a life of poverty), if schizophrenia causes poverty (because it is difficult to hold down a job when suffering the symptoms of this disease), or if some third factor causes both schizophrenia and poverty (because being raised by schizophrenic parents might leave a person vulnerable, independently, to both schizophrenia and poverty).* *Correlation between two factors always implies all three of these causal possibilities, and correlational analysis never allows us to discern which of the three is correct.*

Because strong correlations do not allow causal inferences, Morgan didn't want to jump to the conclusion—as some of his contemporaries did—that different chromosomes *caused* different sexes, or to extrapolate from such a finding and conclude that the chromosomes contained Mendel's "factors" (which by that time were being called "genes"). Instead, he set about doing the experiments necessary to establish that genes were *not* on the chromosomes (in fact, Morgan suspected that genes did not even reside in cells' nuclei—wherein cells' chromosomes reside—but in a completely different part of the cell).

To do his studies, Morgan needed an organism that could survive laboratory life, that would develop quickly, that would reproduce freely and in quantity, and that could be easily anesthetized for examination. He chose to study a common fruit fly, *Drosophila melanogaster*, because it met all his requirements. In addition, *Drosophila* has only four pairs of chromosomes, making these flies relatively simple; moreover, at times during its life cycle, the chromosomes in some of their cells grow to giant proportions, so they are relatively easy to study, even with weak microscopes. Armed with his fruit flies, Morgan set out to demonstrate that genes are not even located in cells' nuclei, let alone on chromosomes. But experiments don't always generate the results we expect; instead, he wound up demonstrating that genes *are* located on chromosomes. Thus, Morgan concluded, chromosomes are

responsible for the development of inherited traits. In 1933, he won the
Nobel Prize in Medicine or Physiology for this groundbreaking work.

Morgan's experiments did not show that genes are the sole efficient
cause of sex (genes couldn't possibly be the sole efficient cause of sex,
because it is possible to have the Y chromosome characteristic of normal
men but the traits of a normal woman—more on this later). Nonetheless,
they generated an enormous amount of excitement in the scientific community. Prior to Morgan's work, genetics and embryology were unified:
early theories of trait origins tried to explain both the transmission of traits
across generations *and* the way in which those traits are formed in development. But with the discovery of actual physical molecules that could be
inherited (that is, transmitted from generation to generation) and that
were also implicated in trait development, a major schism began to grow
within biology.

In showing that genes are on chromosomes, Morgan and his laboratory "established the gene theory [the cornerstone of modern genetics]
and set genetics on a course that diverged from embryology." In the after-
math of Morgan's success, geneticists, who previously had been limited to
conducting Mendel-style crossbreeding studies, began to study chromosomes per se, only studying an animal's specific traits insofar as the presence of a trait was *correlated* with the presence of one or more specific
genes. In contrast, embryologists (or more broadly, developmental biologists) continued to experimentally explore the efficient causes of development. In short order:

> Genetics and development went their separate ways, evolving their own
> techniques, rules of evidence, favorite organisms, journals, vocabulary,
> and paradigmatic experiments. Genetics textbooks stopped discussing
> embryology, and embryology texts ceased discussing genes.

Gottlieb has commented on this cataclysm as well:

> The early pioneers of experimental embryology realized that there
> was a potential for understanding heredity by taking a causal-analytic
> [read: nonteleological] approach to the study of embryonic development.
> Rather than move toward that potential, however, there was instead
> an ever-increasing divergence in the study of heredity and embryological
> development. . . . [After 1910] the geneticists would become uninterested in
> development and most embryologists would forsake the study of heredity.

Before long, discoveries in one domain could not be replicated—or even
studied at all—in the other domain, since the animals typically studied by
geneticists (fruit flies) do not befit the techniques of embryologists, and the
animals typically studied by embryologists (sea urchins and amphibians) do

not befit the techniques of geneticists. Within two decades, researchers working in these subfields could no longer communicate with each other. In fact, the growth of the chasm between them was not even fueled by the impassive, logical arguments usually thought to characterize scientific debate. Instead,

> Hostility between embryology and genetics also emerged. Geneticists believed that the embryologists were old-fashioned and that development would be completely explained as the result of gene expression. . . . Conversely, the embryologists thought the geneticists to be irrelevant and uninformed. . . . The debate became quite vehement.

Thereafter (until very recently),

> genetics and embryology were practiced as independent sciences even though both of them [were involved fundamentally in studying developmental outcomes]. . . . In all of this the gene or genetic material took on a separate existence that made it stand somewhat outside of, or distinct from, the developmental process as such.

For nearly a century now, geneticists have studied fruit flies while embryologists were studying amphibians, geneticists have looked at the correlations between the presence of certain traits and the presence of certain genes while embryologists were doing transplantation experiments, and geneticists have discussed DNA while embryologists were discussing differentiation (all of which we will consider in later chapters). *And all along, both types of scientist have been working on problems related to the same question: Why do we have the traits we do?* But while embryologists (and most scientists who study development at any stage of life) have been looking for the efficient causes of traits in development, geneticists have been asking how traits are passed from generation to generation.

A POTENTIAL SYNTHESIS

The twin inquiries into genetics and development require integration before the great problems of biology and psychology can be solved. While it is an exercise in futility to try to explain development without understanding how genes influence this process, genetics alone cannot explain the origin of traits either; we know from Driesch's work that development is an epigenetic process. The question of how genetically equipotential cells become specialized (for example, as brain cells or heart cells or blood cells) remains unanswerable by genetic analyses that ignore nongenetic influences; only studies of development can answer this question. Thus, a synthesis of genetics and development is required.

The good news is that recent advances in both understanding and technology have allowed such a synthesis to appear on the horizon. According to Scott F. Gilbert:

> We are on the verge of a Renaissance of mammalian developmental genetics. . . . Recombinant-DNA technology has revitalized old embryological questions, has strengthened the ties linking embryology and genetics, and has made possible new studies of gene [activity during early mammal development].

The implications of this synthesis are likely to be monumental.

George F. Michel and Celia L. Moore note that the unification of the biological and behavioral sciences—into an interdisciplinary field called psychobiology—has already produced impressive consequences "in medicine, mental health, and education . . . [and that since these] are basic institutions of society . . . changes in them will ramify throughout the society more generally." While it is too early to predict the consequences of wide societal acceptance of the ideas that would follow from the reunification of genetics and development, this event will probably have implications every bit as profound as those associated with the emergence of psychobiology. Why? In part, because such a reunion will bring with it a rejection of the notion of genetic determinism.

2

WHAT GALTON'S EUGENICS
HAS WROUGHT
Behavior Genetics and Heritability

The famous Galtonian law of regression . . . pretended to have established the laws of "ancestral influences" in mathematical terms. Now . . . these laws of correlation have been put in their right place; such interesting products of mathematical genius may be social statistics *in optima forma*, but they have nothing at all to do with genetics or general biology! Their premises are inadequate for insight into the nature of heredity.

—*W. Johannsen (1911), p. 138*

Two of my friends, who are sisters, tell the story of walking along the beach during sunset, eavesdropping on a conversation in which their children are avidly engaged. Although Sam is only six years old, he is already as tall as his eight-year-old cousin Jake. Commenting on this unusual state of affairs, Sam says, "I think I'm already as tall as you are!" Jake acknowledges this obvious truth, and says "Yeah. I think you're probably gonna be taller than me when we're grown-ups, because your parents are tall and my parents are short—these things are all in the gems."

Jake's lexical confusion notwithstanding, it is apparent that the idea that our genes determine our traits is firmly embedded in the fabric of our culture; how else could children with so few years of experience in the world already believe such things? This idea is such a part of our intellectual inheritance that it currently strikes many adults as intuitively obvious. But such ideas need to be carefully examined; in ages past, we believed the Earth was flat just because it looked that way. Is there similar (ultimately weak) "evidence" behind our intuitions about the origins of our traits? Certainly.

The ordinary lifelong observation that some traits seem to develop independently of our experiences is often taken as "evidence" that our genes must cause these traits. For example, traits that are present at birth, like five fingers on each of our two hands, seem to develop in the *absence* of experience; this contributes to the impression that they must be "genetically determined." Similarly, some traits that are not present at birth—for instance, secondary sex characteristics like facial hair in men—nevertheless appear during adolescence even in the seeming absence of experience with specific environmental events. Some psychological characteristics, too, seem to develop independently of the conditions in which a person is reared. For example, orphaned human babies sometimes develop traits that are reminiscent of their deceased parents, even though the environments in which these babies develop provide no models for these characteristics. Such observations seem to imply that the appearance of certain traits is somehow predetermined—presumably by genes—since these traits do not seem to depend on specific experiences for their development.

In contrast, other traits are very obviously influenced by the ways in which children are nurtured. The specific language that you speak is probably the language that you heard spoken in the environment in which you developed. In many cases, your religious beliefs were shaped by the religious beliefs of your parents. Still other traits seem to be just a *little* impacted by the events of our lives; the extent to which you are shy might have changed somewhat as a result of your experiences, but if you are still relatively shy, you probably feel that nothing could ever turn you into the gregarious life of the party. We are left with a sense that there is a continuum of extents to which traits can be affected by experience. As a result, it seems reasonable to try to determine scientifically the extent to which specific traits can be influenced by environmental factors. Such a project might even involve trying to assign numbers to traits in order to represent their positions on the continuum: 100 percent could be defined to mean "completely open to environmental influence" and 0 percent could be defined to mean "not at all open to environmental influence."

The idea that it is possible to measure how *much* the environment contributes to our characteristics certainly appeals to common sense; when this idea first emerged in the nineteenth century, it appealed to the common sense of scientists as well. Consequently, scientists at the time began trying to develop statistical tools that could measure the extent to which particular traits can be affected by experience. Unfortunately, the work of these scientists reflected their nineteenth-century conceptualizations about the causes of traits, and these ideas have since turned out to be hopelessly simplistic. In the long run, the most important statistical tool to result from this work—the heritability statistic—wound up taking obsolete (and completely untenable) conceptualizations about trait origins and firmly embedding them in our culture's belief systems. If you are under the impression that scientists have shown a certain characteristic to be caused by genetic factors, the ultimate source of this mistaken belief—whether or not you know it—was a study utilizing the logic underlying heritability statistics; these statistics undergird the widespread, but erroneous, belief that genes can determine traits. In fact, even though their name sounds like they should serve as a measure of a trait's "inheritability," heritability statistics do not even reflect the extent to which traits will be "passed down" from parents to their offspring. By and large, then, the public's current confusion about what genes can and cannot do can be traced back to the use—and misinterpretation—of these statistics.

Heritability statistics grew out of the work of one man: Francis Galton. While philosophers predating Galton contemplated the problem of trait origins as well, Galton alone is responsible for showing us the conceptual, methodological, and statistical path that has produced the data underlying the contemporary public's beliefs about trait origins. In fact, the basic conceptualization of the problem and the basic "scientific" method by which the problem is tackled has remained relatively unchanged since Galton first began working on it 130 years ago. And since the scientific underpinnings of our current misconceptions are rooted in Galton's ideas, laying them bare demands first revisiting Galton himself. But be forewarned: the confusion generated by a century of exposure to heritability statistics now runs so deep in people's minds that alleviating it will take us through some territory that might seem downright counterintuitive.

GALTON

The fact that Galton was closely related to one of the most important thinkers of the nineteenth century likely played a role in drawing him to the problem of trait origins; Francis Galton and Charles Darwin were first cousins. Galton's mother, Violetta Darwin, was Charles's aunt.* Galton was aware of his cousin's theory of evolution prior to most of the rest of the world; he later reported in an autobiography that Darwin's *On the Origin of Species* had stimulated his interest in how traits are inherited. In addition, the particular trait that Galton reported on in his first book—titled *Hereditary Genius* and published in 1869—was a trait that he called "eminence"; the fact that his cousin was famous no doubt contributed to his interest in the inheritability of this characteristic.

To study the inheritability of eminence, Galton searched biographical encyclopedias for information on the most distinguished people of eighteenth- and nineteenth-century England. His research revealed that a disproportionately large percentage of these "geniuses"—the country's most respected politicians, judges, lawyers, scientists, military commanders, artists, writers, and musicians—were related to one another by birth. As a result, Galton came to the novel-at-the-time conclusion that "heredity governed not only physical features but also talent and character." How, he then wondered, can we *explain* the observed fact that eminence runs in families?

In contemplating possible explanations, Galton considered the importance of only two broad factors: nature and nurture. Actually, he was the first scientist to use these particular terms,* thereby defining the character of the debate that would ensue over the next century. Of particular significance, the distinction he made between nature and nurture was rather sharp. And since he felt "perfectly justified in attempting to appraise their relative importance" to the appearance of various traits, his subsequent work involved trying to devise scientific and statistical methods that would be suitable for doing just that. Such tools, he thought, would ultimately help him explain why certain traits run in families.

In 1883, with this goal in mind, Galton proposed the use—and outlined the details—of the twin study, a new approach designed to evaluate the relative contributions of nature and nurture to the appearance of traits. Galton believed that the study of twins would allow him to distinguish between "the effects of tendencies received at birth, and of those that were imposed by the special circumstances" of twins' unique lives. He thought that once he was able to distinguish between the effects of nature and nurture, he would then be able to determine the relative importance of each in the formation of the twins' traits.

By studying twins' (or their parents') responses to questionnaires, Galton determined that twins who were similar to one another as youths remained similar to one another as elderly adults, both in body and in mind, even when their later lives had taken them into different environmental circumstances. In addition, he observed that twins who were dissimilar to one another in youth—for instance, twins of different sexes—remained dissimilar in old age, even if they experienced an "identity of nurture." Galton concluded, "Nature is far stronger than Nurture within the limited range that I have been careful to assign the latter."

This result did not surprise Galton at all; in fact, eight years earlier, he had offered the rank speculation that "when nature and nurture compete for supremacy on equal terms . . . the former proves the stronger." Six years before *that*, Galton had jumped to the same conclusion after studying his eminence data. Even though his eminence study—which was *not* a twin study—was not suitable for teasing apart nature and nurture, he concluded nonetheless that eminence runs in families because of nature, not nurture. In interpreting the results of this study, Galton ignored what is now obvious to us: in addition to sharing ancestors, Galton's eminent subjects were all raised in upper-class environments, with all of the educational, nutritional, and material advantages that such environments have always conferred. Clearly, regardless of the potential usefulness or validity of his new twin-study methodology, Galton had a preexisting bias that sometimes prevented him from seeing the contributions that environmental factors can make to the appearance of our traits.

THE GENESIS OF EUGENICS

Galton's conclusion that nature is more important than nurture was not his most notorious assertion. When he found that eminence—like physical traits such as height—ran in families, he quickly concluded that behavioral and personality traits could be artificially "selected," much as a rancher might produce an especially fast horse by selectively breeding a fast stallion with a fast mare. By 1869, Galton had already publicly presented his radical view that "it would be quite practicable to produce a highly gifted race of men by judicious marriages during several consecutive generations." Galton held this conviction even though he lacked any understanding at all of the *mechanism* by which behavioral and personality traits could be inherited.

Thus, after seeing in his 1883 twin data support for his preconceived notions about the primary importance of nature in trait formation, Galton coined the word "eugenics"—meaning "good in birth"—to refer to the "science" of improving humanity by selective breeding of the sort commonly employed on farm animals. Ultimately,

> [He] suggested that the state sponsor competitive examinations in
> hereditary merit, celebrate the blushing winners in public ceremony,
> foster wedded unions among them at Westminster Abbey, and encourage
> by postnatal grants the spawning of numerous eugenically golden
> offspring. (Some years later [Galton] would urge that the state rank
> people by ability and authorize more children to the higher- than
> to the lower-ranking unions.) The unworthy, Galton hoped, would be
> comfortably segregated in monasteries and convents, where they would
> be unable to propagate their kind.

A short 50 years later, this sort of thinking contributed to the brutal social policies of the Nazi regime and ultimately to the extermination of millions of innocent people.

But before the atrocities of the Nazis made plain the inherently evil nature of state-imposed eugenics programs, Galton's proposals didn't strike everyone as being particularly bad ideas. Scientists of the time understood that if Galton's hunch was correct—that behavioral and personality characteristics could be inherited like physical characteristics—this would be an important piece of information. After all, it is only if we understand how our traits are formed that we can conceive of intervening in their formation. And as remains the case today, many people in Galton's time believed that intervening to eradicate certain traits—symptoms of schizophrenia, perhaps, or mental retardation—was a good idea. The only questions are whether or not particular interventions are both effective and morally sound. And because Galton and his followers believed that "what Nature does blindly, slowly, and ruthlessly, man may do providently, quickly, and kindly," eugenics was seen as a reasonable approach to bringing some of society's problems under control. Thus, Galton and his followers set out to make eugenics a true science.

Before long, a new branch of biology called population genetics was born, built in part on Galton's ideas. But while it was a new branch of biology that needed Galton's ideas for its inception, it was Galton's extraordinary contributions to what was then a new branch of mathematics, called statistics, that fundamentally changed how scientists of *all* stripes could make sense of their observations of nature. And since Galton's statistical methods ultimately came to underlie the public's false confidence that some traits are "more genetic" than others, it is important to consider them in some detail.

GALTON'S BRAINCHILD

At the 1884 International Health Exhibition in England, Galton began collecting data on people; before he was done, he had measured the weight, height, arm span, breathing power—basically anything that could be measured—of about nine thousand people, generating a huge data set. Poring over this data set, and paying particular attention to measurements he had obtained from adults *and* their parents, Galton stumbled onto a form of data analysis that he later called a "regression analysis." By 1888, he had figured out that a regression analysis is really just a particular form of a more general type of analysis that he called an analysis of "co-relations"; today, we know this as a correlational analysis.

Correlational analysis generates a statistic—a measurement—that reflects the strength of the relationship between two sets of numbers. For example, since weights and heights can both be represented as sets of numbers, Galton's correlational analysis can be used to evaluate how closely height and weight are related. Because weight ordinarily increases as height increases—taller people usually weigh more than shorter people—these two characteristics are said to be positively correlated. Similarly, because life span ordinarily decreases as cigarette consumption increases—smokers die, on average, at younger ages than do nonsmokers—these two characteristics are said to be negatively correlated. Of course, such an analysis can also reveal if two sets of numbers are completely *un*related; weight and vocabulary size among adults are unrelated to each other, since knowing how heavy an adult is does not even slightly improve your chances of guessing the size of that person's vocabulary.

Galton's initial uses of correlation were not in the service of understanding how genes per se contribute to traits. After all, when Galton was doing his work, the scientific community had not yet rediscovered Gregor Mendel's paper on "heritable factors," and the word "gene" would not be coined for several more years. As a result, Galton used correlational analyses only to address questions about the extent to which traits could be said to "breed true," that is, to be inherited *when environmental factors are held constant*. Farmers are, understandably, interested in the likelihood that breeding animals with particular traits will give rise to offspring possessing those traits; Galton's statistics provide this sort of information.

For example, to see if milk production levels breed true, Galton would measure the amount of milk produced by several cows and the amount of milk produced by their mature female offspring. If he then calculated that there is a high positive correlation between these two measurements, this would mean that cows who produced lots of milk were more likely than less productive cows to have offspring that, when mature, also produced lots of milk. Thus, detecting a correlation effectively allowed Galton to make reasonably accurate *predictions* about the traits likely to appear when the offspring of parents with known traits were allowed to develop *in environments similar to those in which their parents developed*. Before Galton, the statistical tools that would have allowed predictions in these sorts of situations did not exist. As might be obvious, correlational analyses like this one continue to be valuable today; they have useful applications in virtually every branch of contemporary science.

Before long, though, Gregor Mendel's paper on trait inheritance was rediscovered, the word "gene" was coined to refer to his "heritable factors," and Thomas Hunt Morgan demonstrated that genes on chromosomes are both influential in trait development and inheritable—actually passed physically from generation to generation. Thus, early in the twentieth century, Galton's statistical techniques were modified slightly and ultimately applied to the task of trying to determine the extent to which genes contribute to the appearance of traits. Work on this task led to the development of the correlational statistic now known as "heritability." Today, this statistic remains the cornerstone of the branch of psychology called behavior genetics. And since the results of behavior genetics studies inform the decisions of those forging our public policies, it is very important to have an accurate understanding of what heritability means.

SEEING DOUBLE

A heritability estimate for a trait is a number that tells the extent to which differences in the appearance of the trait across several people can be "accounted for" by differences in their genes;* computing these statistics is the aim of behavior geneticists. For example, the most widely publicized behavior genetics report of the 1990s was T. J. Bouchard and colleagues' finding that the heritability of IQ is .70. Given the definition of heritability, this means that 70 percent of the differences in IQ found among Bouchard's subjects could be "accounted for" by differences in their genes. But what exactly does that mean?

The best way to understand heritability estimates is to examine how they are produced in the first place. Heritability estimates are usually generated by analyzing data collected in studies of twins.* Twins are important in the study of heritability because women can bear at least two different kinds—so-called identical twins and fraternal twins—and because the two types differ in the extent to which they have the same genes. The genetic differences between twin types result from events that occur around the time of their conception. Fraternal twins are conceived when two of a mother's unique eggs are fertilized at about the same time, each by a different, unique sperm. This gives rise to two unique zygotes, each of which ultimately develops into a unique person. Such twins are typically as different from each other as would be any two zygotes conceived by these parents; that is, they are as similar to—and as different from— each other as are regular siblings. Thus, the only difference between fraternal twins and an ordinary brother-sister pair is that fraternal twins begin their lives around the same time, whereas the lives of ordinary siblings usually begin a year or more apart. In contrast to fraternal twins, identical twins are produced when a single embryo splits in two, usually for unknown reasons. The resulting two embryos are made of cells that contain the exact same genetic material, and they normally develop into people that are unusually similar to each other in certain respects (for example, in their appearances).

The logic of twin studies rests on the fact that identical twins have identical genes, whereas fraternal twins do not. By producing two different types of twins, some with identical genes and some with genetic differences, nature has given us the raw materials for an "experiment"; if we find that both members of an identical twin pair almost always share a particular trait but that both members of a fraternal twin pair do not, then the pattern of variation in the trait can be "accounted for" by the pattern of variation in the genes. For example, if—as appears to be the case—identical twins usually have similar IQs but fraternal twins sometimes do not, the differences among the fraternal twins must have developed because of *something* that was different for them but that was not different for the identical twins. And since fraternal twins have differences in their genes and identical twins do not, it is quite plausible that the "something" in question could be their genetic constitutions. In this way, behavior geneticists can sometimes trace patterns of variation in twins' traits to the patterns of genetic variation that characterize different twin types. Thus, if a behavior geneticist reports that IQ is highly heritable, what he means is that, by and large, individuals with the same genes have similar IQs and individuals with different genes have relatively different IQs.

The logic behind this sort of study is quite compelling; differences among fraternal twins that are *not* seen among identical twins *must* result, somehow, from differences present in the fraternal, but not the identical, twins' genes.* So, what's the problem? The problems appear when we consider what heritability estimates do *not* mean. From this perspective, heritability statistics can be seen to be extraordinarily misleading.

HERITABILITY: AN IMPOSING ILLUSION

Reporting on an analysis of the heights of fraternal and identical twins, Robert Plomin—one of the leading behavior geneticists in America today—writes that the

> results indicate significant genetic effects. For these height data,
> heritability is estimated as 90 percent. This estimate . . . indicates that, of
> the differences among individuals in height in the populations sampled,
> most of the differences are due to genetic rather than environmental
> differences among individuals.

Most people, upon encountering these words, would quite reasonably conclude that genetic factors must be more important than environmental factors in causing a person's height. After all, if genetic factors account for 90 percent of the differences in people's heights, it seems that environmental factors must not account for any more than 10 percent of this variation. And given this (mis)understanding, it would seem that it is probably difficult to influence highly heritable traits with manipulations of the environment. Thus, Plomin's statements, by seeming to suggest that people's heights are largely unaffected by their developmental environments, encourage us to conclude that people's heights are determined *prior to development*, by genes inherited from their parents.

Unfortunately, the fact of the matter is that these conclusions are most definitely not valid. Given the misleading language and concepts of behavior genetics, of course, one could be forgiven for coming to such erroneous conclusions. Nonetheless, the fact remains that heritability estimates are *not* measures of the importance of genes in the production of a person's traits, or of a trait's "openness" to environmental influence.

Heritability: Not about the Relative Importance of Genes

Heritability statistics do not reflect the relative importance of genes in causing traits, because—and this cannot be overemphasized—**heritability estimates tell us about what causes *variation* in traits; they tell us nothing at all about what causes traits themselves.** This difference seems incredibly subtle at first; if the difference between my height and your height can be traced to our differing genes, it certainly *seems* like this means that our respective genes determine our respective heights. I have discussed the difference between "accounting for variation" and "explaining causation" with many extraordinarily smart people who have never encountered it before, and when they first do encounter it, they are typically mystified; you're in good company if you feel the same way. But once you have given the issue plenty of thought, the difference seems about as subtle as a brick.

Consider a simple analogy drawn from nature: the formation of snowflakes.* Snowflakes are formed only in the simultaneous presence of two factors, namely a temperature below 32 degrees Fahrenheit and a relative humidity high enough to allow for precipitation. Now, if on a given day, humidity is high at the North Pole but low at the South Pole, snow will fall only at the North Pole; *in this case, the variation in snowfall across the two locales can be accounted for completely by variation in relative humidity*. But such a circumstance certainly cannot be taken to mean that temperature is unimportant in causing snow. Thus, "accounting for variation" and "explaining causation" are profoundly different from each other. Humidity differences alone are enough to account for the differing snowfalls at the two poles in this example, but only because it is *always* cold enough for snow at both poles, not because coldness is unimportant in causing snow. Whenever a factor does not vary across situations (as temperature does not in the current example), it cannot "account" for variation in outcomes across those situations; still, this does not mean that the factor plays no role in causing the outcomes themselves.

This analogy illustrates how little an "account of variation" can tell us about causation. Snow is *caused* by two factors, even if all of the variation in its presence can be "accounted for" by variation in only one of those two factors. The same holds for our traits; it is quite possible, for instance, for genetic factors to "account" for 90 percent of the *differences* seen in people's heights, *without* genetic factors being any more important than environmental factors in *causing* people's heights. *Regardless* of the heritability of height, a person's height is *caused* by both genetic and environmental factors. Twenty-five years ago, Richard Lewontin offered the following illustration of this point.

Lewontin asked us to imagine planting ordinary, genetically diverse seeds and then letting them grow to maturity in two different environments. Imagine that, within each environment, light, water, and nutrients are uniformly distributed but that one of the environments (environment A) provides its seeds with sufficient light, water, and nutrients, while the other environment (environment B) provides its seeds with just barely enough light, water, and nutrients to survive. At maturity, *all* of the height variation seen among the plants grown in environment A must be due to the genetic diversity that was originally present in the seeds; this *must* be the case, because all the seeds developed into mature plants in the *identical environment* (A). Thus, *none* of the variability in height can be accounted for by environmental variation (since there was none), and heritability of height in environment A must be 100 percent. Similarly, heritability of height in the deficient environment B is also 100 percent, since the height variation among the mature plants grown in this environment is also entirely attributable to the genetic diversity present in the seeds (again, because they all developed in the same environment). Thus, regardless of the environment in which the plants grew, the heritability of height is 100 percent in this example. If we didn't know any better, of course, a high heritability like this might have led us to think that plant height is 100 percent determined by genes. But examining the plants would tell a different story: even third graders know that plants grown in the deficient environment B can be counted on to be shorter, on average, than plants grown in the normal environment A. Thus, even if the heritability of a trait is very high—*in fact, even if it is 100 percent*—numerous *environmental* factors can nonetheless have overwhelmingly powerful effects on the trait's final appearance.

The converse of this story holds as well: a trait with a heritability of zero *cannot* be assumed to be *un*affected by genetic factors. Imagine taking one of the seeds from Lewontin's example and cloning it so as to produce a handful of genetically identical seeds. If we scattered these seeds in a variety of different soil types and let them develop, this time we would find *all* of the variation in the plants' heights to be accounted for by variation in the soils (since there is no genetic variation among the plants at all). In this case, then, the heritability of height would be calculated to be 0 percent.

But surely no one would argue that in this example genes have nothing to do with plants' heights! How could the height of a particular plant be *una*ffected by the plant's genes, when genes participate in constructing the raw materials out of which plants are made? In fact, a trait *can* be importantly affected by genetic factors *even if it is not heritable at all*.

But if *perfectly* heritable traits are consequentially affected by both environmental and genetic factors, and if *non*heritable traits, likewise, are consequentially affected by both of these same factors, then what is the difference between heritable and nonheritable traits? In fact, *there is no difference at all*, at least, not in the extent to which heritable and nonheritable traits can be influenced by genetic and environmental factors.

To see this clearly, consider what it would mean if the heritability of hairstyles was .01, or 90 times less than the reported heritability of height. Is a heritable trait like height more influenced by genetic factors than is a nonheritable trait like hairstyle? No. In fact, *heritability estimates for height can be many times greater than heritability estimates for hairstyles, but this alone would* not *mean that a woman's height is more influenced by genetic factors than is her hairstyle*. Why not? Because heritabilities aside, genetic factors can profoundly influence the development of hair characteristics—curliness, for example—that can influence hairstyle decisions. By the same token, lower heritability estimates for hairstyles than for heights cannot be taken to mean that environmental factors impact hairstyles more than they impact heights; the height a woman attains in adulthood is affected by factors in her developmental environment—her diet, for example—just as the hairstyle she eventually dons is affected by environmental factors such as the advertising she is exposed to. Thus, although this might all seem extraordinarily counterintuitive, the fact is that heritable traits can be just as affected by environmental factors as can nonheritable traits, and nonheritable traits can be just as affected by genetic factors as can heritable traits. The disparity between what we assume heritability estimates tell us (given their name) and what they *really* tell us is *so* great that it is hard, at first, to make sense of it all. But this is true: since heritability estimates account for variation *and do not explain causation*, they tell us *nothing at all* about how genetic and environmental factors influence the development of our traits.

There are important practical consequences associated with the fact that heritability estimates do not tell us about what causes—or can affect—individuals' traits. For example, consider the following erroneous argument that makes use of Bouchard's high estimate of the heritability of IQ. Perhaps, some might argue, we should not bother devoting public resources to school programs designed to raise IQ scores, because the high heritability of IQ means that IQ is largely "genetic," and so not particularly open to environmental influence. The problem with this argument is that nothing could be farther from the truth. Since environmental manipulations can *profoundly* influence the development of traits that are even *more* heritable (even maximally heritable), heritability estimates cannot be appropriately used in this way.*

Heritability: The Exquisite Specificity

At this point, you might find yourself saying, "Wait a minute! If the heritability of a trait tells us nothing about the extent to which genetic or environmental factors contribute to the trait's appearance, what use is knowledge of heritability?" A behavior geneticist might respond that heritability estimates do, at least, help us understand the source of *differences* among people. But even this claim is overstated, since it turns out that *heritability estimates cannot be generalized to situations different from the situation originally studied*. Instead, these statistics tell us only about the causes of differences among people "*in a particular population at a particular time*" and in *particular circumstances.**

Heritability estimates cannot be generalized because as soon as we start changing situations, the whole story changes, and the factors that once accounted for different outcomes might no longer do so. Consider again the formation of snowflakes. When we look at the causes of variation in snowfall at the poles, we find that the variation is accounted for by variation in relative humidity alone. But if we study variation in snowfall at several locations in an extremely humid Costa Rican rainforest, we will find that snowfall variation will be accounted for by variation in *temperature* alone. (Temperature varies in Costa Rica as a function of altitude; even in countries near the equator, snow falls on high mountaintops.) Thus, what "causes a difference" under one set of circumstances might not account for *any* of the variation detected in a different set of circumstances. As a result, accepting that a particular factor is a "cause" of a difference requires us to hold constant *every other factor* in a situation.

Heritability estimates must be understood in this same way. If *everybody* developed in environments just like those in which Bouchard's subjects developed, perhaps then we would be justified in thinking that a high heritability for IQ means that IQ differences are invariably accounted for by differences in genes. But we do *not* all develop in similar environments, and it remains likely that genetic differences that account for IQ variation in some environments do not account for IQ variation in others.

What is the value, then, of knowing what accounts for *differences* among people *in a unique situation*, if it is within our power to easily change situations? Understood this way, it is not clear that it is at all useful to know what accounts for differences. Knowing how a full complement of influential factors collectively *causes* individuals' traits is the more important goal, by far. For this reason, I focus in this book exclusively on questions about the *causes* of traits, and not on questions about the sources of differences among individuals.

Heritability: Unrelated to "Inheritability"

The upshot of this situation is that even though "heritability" sounds like it would reflect how "inheritable" a trait is, in a world of changeable environments, heritability estimates do not, in fact, do any such thing.* The truth is that the statistical concept of heritability bears almost no resemblance whatsoever to our intuitive concept of "inheritability."

Consider first a trait that seems like it is part of our "biological inheritance," namely the presence of five fingers on each of our hands. Given our confidence that we inherit this trait, most of us would probably expect this trait to be highly heritable. But as the philosopher Ned Block has explained, most of us would be mistaken:

> The *heritability* of number of fingers and toes in humans is almost certainly very *low*. What's going on? If you look at cases of unusual numbers of fingers and toes, you find that most of the variation is environmentally caused, often by problems in fetal development. For example, when pregnant women took thalidomide [a drug later implicated in the production of birth defects] . . . many [of their] babies had fewer than five fingers and toes. And if we look at numbers of fingers and toes in adults, we find many missing digits as a result of accidents.

But if most of the variation in people's digit numbers can be traced to variation in environmental factors, then little of the variation in digit number is "accounted for" by genetic variations; in fact, human genes only rarely—if ever—contribute to the appearance of just four fingers on a hand. Thus, heritability of this trait would be calculated to be rather low. For similar reasons, many common traits—possessing opposable thumbs, four limbs, or teeth located inside your mouth—are not particularly heritable, even though it is inconceivable to most of us that these traits are not "inherited." Is it just a minor inconvenience that "heritability" and "inheritability" sound so much alike when in fact they mean very different things? I think not; this terminology leaves most of us quite confused.

Consider the opposite side of the same coin: What do heritability estimates tell us about traits that we intuitively sense are *not* inheritable? Block addressed this situation as well, writing "Some years ago when only women wore earrings [in America], the heritability of having an earring was high." It seems as if this can't be right; we all know that earring

wearing is *very* affected by cultural factors, and so is unlikely to be "inherited" according to any traditional definition of this word. Nonetheless,
earring-wearing behavior in 1950s America was, in fact, highly heritable.
What's going on here?

This counterintuitive finding reflects the fact that in mid-century
America, there actually were certain genes that were found in most earring
wearers and that were hardly ever found in non–earring wearers. Think
about that for a second. Does this mean that there actually are genes "for"
earring wearing? Not at all, *because correlation tells us nothing at all about causation;* genes whose presence is correlated with earring wearing need not
play *any direct role* at all in causing that behavior. As it happens, the genetic
factors consistently found in earring wearers in the 1950s were two X
chromosomes, which characterize almost all women and very few men.
But X chromosomes do not *cause* earring wearing any more than do ovaries
(the presence of which were also highly correlated with earring wearing in
the 1950s). Instead, since only women wore earrings at that time—*for cultural reasons*—there was a high correlation between the presence of two X
chromosomes and the presence of jewelry dangling from the ears.

Amazingly, the discovery that specific genes are highly correlated with
the appearance of a trait is enough for behavior geneticists to conclude
that the trait is heritable, *even if the appearance of the trait obviously depends
on cultural factors*. Why is that? Because for behavior geneticists, the goal
is to predict the appearance of traits by looking at genes, and if there is a
high correlation between the presence of some genes—any genes!—and
the presence of a trait, then looking at the genes allows them to make predictions about the trait's appearance. For behavior geneticists, the fact
that X chromosomes do not directly *cause* earring wearing or that cultural
factors *do* have a direct causal role in producing this behavior are simply
not of interest. As far as they are concerned, if there is a *correlation*
between earring wearing and the presence of certain genes—however
*in*directly related to earring wearing those genes are—then earring wearing is heritable, even if environmental factors play essential roles in causing the trait. Knowing what effects nongenetic factors have on a trait's
appearance is simply not of interest to behavior geneticists, who consequently just ignore these factors.

Take note: this is not just an abstract argument without implications. In
fact, it would be entirely possible using behavior geneticists' correlational
methods to find that genetic variations among American blacks and whites
"account for" most of the variation in their IQ scores, *but this would not mean
that the IQ differences are caused by the genetic differences*. Instead, as Block's
earring example shows, a finding that IQ is heritable could simply mean
that people with genes that contribute to different skin colors develop dif-

ferent IQs because of how society treats people with different skin colors. In the language of behavior genetics, genes that contribute to skin color differences could fully "account for" IQ differences, even if these genes influence IQ *only* via racist attitudes and behaviors present in our society.*

GALTON'S LEGACY

So where does all of this leave us? First, heritability estimates tell us nothing at all about what causes an individual's traits. Second, heritability estimates do not reflect the extent to which traits are impervious to environmental influences. Third, heritability estimates apply only to people in a particular population, at a particular time, and in particular circumstances; they can *never* be appropriately generalized to other populations. Finally, despite its evocative name, heritability estimates do *not* tell us how likely it is that parental traits will be *inherited* by their offspring. Instead, heritability estimates are capable of indicating—counterintuitively—that characteristics like finger number are not heritable and that characteristics like earring-wearing behavior are.

Ultimately, then, heritability estimates are not at all what they appear to be. If, in the future, you hear that a specific trait—let's take alcoholism, for example—is highly heritable, try to resist the temptation to immediately think several things that you might otherwise think. First, even if alcoholism were *perfectly* heritable, this would not mean that it will necessarily afflict the children of alcoholics; even *perfectly* heritable traits are affected by the environment, so raising the children of alcoholics in an environment different from the one in which their alcoholic parents were raised could make all the difference. Second, if alcoholism is heritable, this does not mean that it is caused by genetic factors; it needn't even necessarily be the case that genetic *differences* between alcoholic and nonalcoholic individuals *directly* contribute to the differences in their drinking patterns. And, finally, if alcoholism is heritable among Iowans, it need not be the case that it is heritable among Ohioans. (I am not simply being recalcitrant here: heritability estimates calculated for one population *do not apply* to another population.) These are truths that are widely misunderstood by a public that nonetheless avidly consumes heritability data.

Given these problems, is there some *other* measure that better captures our intuitive notion of "inheritability?" Unfortunately, there is not. The heritability estimate, with all of its inescapable problems, is the statistic that comes closer than any other to capturing this intuitive notion. In fact, *the scientific data that underlie the public's misperception that some traits are more "in the genes" than others are those that come from heritability studies; there are scant, if any, other scientifically collected data underlying this conviction.* Individuals certainly have anecdotal "evidence" behind their intuitions that not all traits are equally inheritable, but in the past 100 years, the main scientific approach to studying inheritance has been the one pioneered by Galton.* Unfortunately, the foregoing analysis of heritability requires us to view the endeavors and claims of Galton's followers with skepticism. Heritability estimates are misleading at best, and the interpretations they encourage are simply wrongheaded at worst; they do not illuminate the origins of our traits as we might otherwise have assumed. P. R. Billings and colleagues agreed with this assessment when their analysis of the twin study literature led them to conclude that "identical twin studies offer no convincing evidence of the genetic basis of human behavior."*

Actually, the situation is even worse than it appears. I have been describing how the logic behind *an idealized, perfect* twin study would still limit how we could properly interpret the study's results. Lamentably, there are other logical problems that bedevil *actual* twin studies conducted in the real world.

3

LISTENING TO TWINS
A Critique of Twin Studies

Among the more famous identical twins ever to be studied by behavior geneticists are Daphne Goodship and Barbara Herbert, who were born in 1939 to a single mother of Finnish descent. The twins were tragically orphaned shortly after their birth when their mother committed suicide; the baby girls were subsequently adopted into separate homes, one going to a middle-class family living in a small town north of London, the other going to a slightly poorer family in the city itself. For the next 39 years, Daphne and Barbara lived completely separate lives, finally meeting each other for the first time at London's King's Cross Station in May 1979.

By all accounts, the twins arrived at the meeting dressed in similar colored dresses and brown velvet jackets. In their ensuing meeting, they quickly discovered that they shared some truly remarkable similarities: both women had a crooked pinkie finger, had dropped out of school at age 14, hurt their ankles falling down the stairs at age 15, and met their husbands at a local dance at age 16. But the similarities did not stop there: the women's pregnancy histories were identical, with both women first experiencing a miscarriage and subsequently giving birth to two sons followed by a daughter. Neither woman had ever voted except for once—each woman voted in the one year when she worked as a clerk at the polls. Possibly most amazing, both women had developed the odd habit of pushing up her nose with the palm of her hand, a habit that each had given the name "squidging." Can we conclude that their identical genes account for these trait similarities? In the absence of an enormous amount of thought about twins, it is extremely easy to draw such erroneous conclusions, particularly in the face of observations that sometimes astonish us.

IDENTICAL TRAITS: COINCIDENCE?

How are we to understand the origin of the similarities between Jim Springer and Jim Lewis, identical twins separated at birth who, upon being reunited 39 years later, discovered that in addition to sharing the same name, they had both twice chosen spouses with the same name, worked part-time in law enforcement, and smoked Salem cigarettes? The events recounted in such stories typically strike us as too incredible to be attributable to coincidence. Nonetheless, some of these events *might*, in fact, best be understood this way.

To evaluate the possibility that a particular similarity among identical twins is really just coincidental, we would need to determine just how common the trait in question is. For example, how likely is it that two randomly selected men would coincidentally have the same name? The answer to this question depends on how common the particular name is. A century ago, fully half of all children born in North America were given one of the ten most popular names for their sex; under these circumstances, two randomly chosen individuals of the same sex would not be all that unlikely to have the same name. While the penchant for giving babies popular names is not now quite what it once was, it remains the case that if we queried enough pairs of randomly selected (and so completely unrelated) contemporary American men, we should *expect* some of them to have wives sharing the same name. By the same token, to properly evaluate the likelihood of reared-apart twins preferring, for example, the same cigarettes, we would want to know things like how many different cigarette brands are available in the twins' hometowns or how many reared-apart twins do *not* smoke the same cigarettes. But because a discovery that identical twins reared apart do *not* smoke the same cigarettes is anything but newsworthy, the public rarely encounters comparative data of this sort. As a result, we wind up with the illusory impression that most identical twins possess incredibly similar traits.

In truth, so-called identical twins are *never* identical. But such twins likely share *some* very similar traits, and some of these matches probably cannot be attributed to coincidence; the odds of Daphne and Barbara both developing a tendency to nose-squidge seem vanishingly small. So we are left with the question: Do twins' similarities on such unusual traits reflect the fact that they have identical genes?

Not necessarily. True, if "identical" twins each have a certain distinctive trait despite being raised in *completely* different environments, then the only possible source of the traits' similarity *would* be the twins' identical genes.* But to correctly attribute similarities in twins' traits to the twins' genetic similarities, we must be sure that the trait similarities did *not* appear as a result of *environmental* similarities. And it turns out that this is not as easy to ensure as it might appear to be: even when twins are separated at birth, they are often reared in environments that share many similar—and important—characteristics.

RAISING TWINS IN "DIFFERENT" ENVIRONMENTS

A glaring form of this problem plagued early studies of twins reared apart, enough so that it earned a formal name: *selective placement*. The problem of selective placement arises from the fact that adoption agencies have traditionally tried to place adoptees into homes that are not very different from the homes they would have had, had they remained with their biological parents. In particular, when considering the appropriateness of a potential placement, "adoption agencies will consider such factors as socioeconomic status, religion, and cultural interests." As a result, in earlier twin studies,

> many of the twins who were "reared apart" were often raised by close relatives, lived in the same or similar neighborhoods, and even attended the same schools. Moreover, when twins did not live with relatives, the adoption agencies, in keeping with standard policy, placed them in home environments similar to their original family environments.

Thus, even if twins in these studies were reared apart, it is definitely *not* the case that they were reared in *very* different environments (let alone *completely* different environments). And, as a result, any similarities in their traits could conceivably be accounted for by the similarities in their rearing environments.

In one early study of twins reared apart, for example, "the most common pattern was for the biological mother to rear one of the twins,with the other twin being reared by the maternal grandmother or by an aunt." This is obviously problematic, as the environments that two sisters—a twin's mother and aunt, respectively—provide for their children can be expected to be rather similar, on average. The same sort of methodological flaw tainted the results of case studies conducted in 1937 by H. H. Newman, F. N. Freeman, and K. J. Holzinger as well. In this research, Kenneth and Jerry, one of the 19 twin pairs who were studied,

> had been adopted by two different families. Kenneth's foster father was "a city fireman with a very limited education." Jerry's foster father, by contrast, was "a city fireman with only fourth-grade education." . . . Harold and Holden, another pair studied by Newman et al., had each been adopted by a family relative. They lived three miles apart and attended the same school.

Thus, even though these twins—like Daphne and Barbara, who both grew up in and around London—were reared apart, their "different" developmental environments were actually quite similar. Consequently, their similar traits could have resulted from similarities in their genes, similarities in their environments, or, most likely, from similarities in the gene-environment interactions that occurred as they developed.

As it happens, this problem is surprisingly intractable. While researchers can take care to ensure that they refrain from studying twins who were adopted at birth into *very* similar homes, the fact is that twins are never adopted into truly different environments. Even if one twin is adopted into the family of a wealthy New York lawyer while the other is adopted into the family of a struggling Kansas farmer, there are still myriad ubiquitous features of their cultural environments—newspapers, Barbie dolls, TV movies, Burger King—that are more likely than not to characterize *both* twins' adoptive homes. The existence of such ubiquitous cultural features makes it virtually impossible to ensure that developing twins experience *truly* different environments as they grow up.*

In fact, it turns out that it is actually *impossible to raise two children in* completely *different environments, because some features characterize* all *environments in which human beings develop*, even wildly disparate environments like Manhattan and Mongolia. Examples of such ubiquitous features include things like the presence of terrestrial gravity, oxygen, water, nutrients, a certain range of temperatures, and, except in extraordinary situations, communicative and caregiving adults. Such factors, which are known to influence the development of our traits, cannot be ignored without jeopardizing our understanding.* Their very ubiquity actually makes them hard for us to notice—in fact, we are liable to forget about them if we aren't careful—but they ought not be ignored simply because they are ubiquitous. It is precisely *because* these features will characterize *any* two environments into which twins can be adopted that we can *never* completely rule out the possibility that a specific twin-pair's similarities can be accounted for, in part, by similarities in their developmental environments. Ordinarily, when we refer to identical twins reared apart, we *imply* that they were reared in *different environments;* but while some features of reared-apart twins' adoptive environments are likely different, others most certainly are not.

What this means is that if you hear about identical twins separated at birth, who, upon first meeting each other as adults, discover that they are both fond of green hats, that they both have IQs near 125, and that they both vacation in Kalamazoo every spring, you *cannot* infer that these similarities reflect genetic influences. Given the problems I have been describing, anecdotal data like these can *never* be used to draw such conclusions. Instead, such similarities *could* reflect genetic similarities, but they could also reflect coincidence, environmental similarities born of selective placement, environmental similarities born of the fact that we all develop in environments that share certain features, or similarities in the interactions that occur during development between individuals' genes and their environments.*

Similar Environments, Similar Twins: The Equal Environments Assumption

I recently interviewed my friend's 10-year-old sons about their lives as "identical" twins. Steven and Daniel are really lovable; they're also extraordinarily precocious and even more hyperarticulate than the following passages suggest. Their large vocabulary probably owes something to the fact that their mother is a physician. How else might one explain Steven's favorite word?

> DM: *Tell me, when you get older, do you think you're gonna be more alike or less alike, or about the same as you are now?*
> Steven: *Less alike.*
> Daniel: *I think we'll be less alike, 'cuz I'll keep eating the same way I do and he'll keep eating the same way he does . . .*
> DM: *So you think Steven's gonna get kind of big and you're gonna be kind of . . .*
> Steven: *He's gonna be a twig. . . . I swear I'll just step on him (pretends to step, and gives out a little yelp). "You broke my spleen!"*
> DM: *(Laughing) You can't break a spleen!*
> Steven: *That's my favorite word: SPLEEEEEN!*

Toward the end of the interview, I tried to get past the boys' playful competitiveness, to see if I might be able to find a little of that storied rapport that twins are purported to possess in spades. I asked:

> DM: *Do you guys have words that you say to each other that only you guys understand?*
> Daniel: *Yes, sometimes.*
> Steven: *Sometimes. Like . . .*
> Daniel: *We have like, we make up languages that no one can understand. Even we don't really understand it, we just say it because it's fun.*
> Steven: *And it's weird, you know? That's how it works.*
> Daniel: *That's how it works!*
> *(Laughter all around)*

Even though my search for a special bond of communication between Daniel and Steven was thus foiled, when I asked them about other kinds of twins, it was clear that they understood that there was something special about being "identicals":

> DM: *Do you know any nonidentical twins?*
> Steven: Yes, I do. I know a pair of fraternal twins. They are one of my best friend's older brother and sister . . .
> DM: *How old are they?*
> Daniel: *They're going into the seventh grade.*
> DM: *OK, so they're a little bit older.*

Steven: *A lot older!*
DM: *Do you think . . . um . . .*
Steven: *I could get you an interview with them?*
DM: *No, I'm more interested in how you think they're . . .*
Daniel: *Different from us?*
DM: *Or if you think they are. Do you think there's anything different about being that kind of twin as opposed to being the kind of twins you guys are?*
Steven: *Yes, 'cuz it's easier to tell who's who, and you can diff, diffee-enterate, diffee . . .*
DM: *Differ . . .*
Steven: *. . . differentiate—(grinning) can't . . . even . . . talk—you can differentiate the way they are by sex.*
Daniel: *Steven, never drink beer. You already have trouble speaking.*
Steven: *(Backhands Daniel's arm, smiling)*

Without knowing it, Steven—who, in my opinion, deserves kudos for even *attempting* to come out with "differentiate" as a 10-year-old—thereby tapped into one of the thornier problems that plague modern twin studies. Contemporary twin researchers' methodologies allow them to get around the problems of coincidence or the existence of ubiquitous features in human developmental environments, problems that prevent clean interpretation of anecdotes about identical twins reared apart. But even the most well-conceived and well-conducted of these researchers' studies are undermined by the fact that their "experimental" methods require them to make some assumptions that are probably unfounded. In particular, current twin study methodologies require researchers to assume that different kinds of twins develop in equally similar environments.

Why do twin researchers have to assume that the developmental environments of identical and fraternal twins are *equally similar* (or equally different, which, when you think about it, amounts to the same thing)? To understand why this is so, imagine two pairs of twins who are adopted into four different homes. Frick and Frack, our first pair, are "identical" twins who are adopted at birth into homes in suburban Philadelphia and suburban Baltimore, respectively. Ralph and Maryanne, our other pair, are fraternal twins (obviously, since they are of difference sexes); they are adopted at birth into homes in Baltimore and New Delhi, India, respectively. In this case, if Ralph and Maryanne do not ultimately share traits that Frick and Frack do share, their differences could be due to differences in their rearing environments just as easily as they could be due to differences in their genes. Thus, we can only attribute trait differences between fraternal twins to their *genetic* differences if the environments in which fraternal and identical twins develop are *equally* similar. Behavior geneticists are aware that the logic underlying their method *requires them to assume* that the developmental environments of fraternal and identical twins are equally similar; they have even given it a name, calling it "the equal environments assumption." Unfortunately, the need to make this assumption threatens the validity of their studies; there are good reasons to believe that the environments of fraternal and identical twins are *not* equally similar.

Why is there reason to suspect that fraternal and identical twins are *not* raised in equally similar environments? The argument goes like this: first, imagine that Frick and Frack are both beautiful enough to appear in diaper commercials. Unfortunately, Ralph and Maryanne, differing as much as do ordinary siblings, are not equally beautiful; for her part, Maryanne is diaper-commercial beautiful, but her brother is actually rather homely. If these four infants are adopted and then raised in randomly selected, but otherwise similar, "normal" American homes (let's put them all in Baltimore this time), can we expect their environments to be equally similar? Probably not. The three beautiful babies are likely to encounter throughout their lives people who treat them relatively well; people will seek their company to be in the presence of their beauty, they will grow accustomed to compliments, and they are liable to develop sophisticated social skills as a result of their experiences.* In contrast, poor Ralph is likely to suffer the indignities that the world visits on ugly people, as adults pay him less positive attention and as his peers choose other, aesthetically more appealing children as their friends. The important point here is that *regardless of the apparent similarities of their adoptive environments*, the warm environments *experienced* by Frick and Frack will probably be *more* similar to one another than Maryanne's warm environment will be to the rather less friendly environment that Ralph experiences. And such a difference in similarities is important, given that the logic of twin studies *requires* fraternal and identical twins to be raised in *equally* similar environments.

The eminent plausibility of this scenario suggests that the equal environments assumption underlying twin studies is probably unwarranted. As Michel and Moore note:

> People use physical appearance as a basis for their behavior toward others. The way that people are treated by others contributes importantly to the development of personality characteristics. Because identical twins are physically more similar than nonidentical twins, they will be treated more similarly even when reared apart.

Thus, identical twins—*simply by virtue of being physically more similar than fraternal twins*—probably grow up in environments that are more similar than those in which fraternal twins grow up. Note that this problem cannot be resolved *even when experimenters go to great lengths to try to ensure that all twin pairs are raised in equally similar environments.* In fact, the most heroic efforts of the most competent experimenters will not neutralize this problem; it grows from characteristics that *typify* the different types of twins who are studied in properly conducted twin studies.*

THE EQUAL *PRENATAL* ENVIRONMENTS ASSUMPTION

As it happens, at least one gross violation of the equal environments assumption haunts nearly *all* twin studies, rendering each of them seriously flawed. This violation often goes undetected, though, because twin researchers (and the public as well) commonly assume that the environment influences traits only after birth. As a result, the question of whether or not identical and fraternal twins have equally different *prenatal* environments is often not considered at all. This question is quite important, though, because a variety of our traits develop before we are born.

It seems unlikely, at first glance, that prenatal environments vary enough to be responsible for trait variations; on an initial look, fetuses appear to be effectively isolated from the broader environment, sheltered in their mother's womb. It is now clear, though, that while a uterus is a remarkably controlled and buffered environment for a fetus, it is still variable in important ways. This has become apparent as scientists have discovered how the presence of certain substances in utero—alcohol, for example—can affect the development of a fetus's traits. As a result, it is very important to consider whether or not identical and fraternal twins have *equally* different *prenatal* environments. If twins can be *born* with trait differences traceable to differing prenatal environments, this would undermine the argument that differences seen among fraternal, but not identical, twins are due to differences in their genes. In such a case, the differences between fraternal twins could reflect variations in their prenatal environments.

Are the prenatal environments of fraternal twins *more* different from one another than are the prenatal environments of identical twins? Absolutely. In fact, the prenatal environments of identical and fraternal twins are *not* similar enough to warrant extending the equal environments assumption to the prenatal period. Ironically, though, the data available to address this question have not come from studies of fraternal twins' prenatal environments themselves. Instead, studies of *identical* twins alone have made it clear that the prenatal environments of different types of twins can vary, and that these variations can profoundly influence the appearance of both biological and psychological traits.

Understanding these data requires knowing about a remarkable phenomenon that occurs during the prenatal development of human twins. Under normal circumstances, some of the cells of one-week-old human embryos begin to develop into a membrane called a "chorion" (pronounced kor-ee-on); ultimately, the chorion forms the embryonic part of the placenta, the organ that allows the fetus to share nutrients, oxygen, and wastes with its mother.* Most identical human twins—about two-thirds of them—are formed when a single embryo splits in two sometime between five and nine days after fertilization; since a chorion has developed by this point, both twins wind up sharing a single placenta. In contrast, about one-third of identical human twins are formed when a single embryo splits in two *prior* to five days after fertilization. In this case, no chorion has been formed at the time of the split; consequently, each twin develops its own chorion, ultimately co-constructing its own placenta with the mother. As it turns out, these different prenatal conditions appear to make important contributions to later developing trait differences.

More than 20 years ago, M. Melnick, N. C. Myrianthopoulos, and J. C. Christian reported that among a population of white Americans, identical twins who shared a placenta in utero were significantly more likely to have similar IQs than were identical twins who each developed their own, individual chorion. Reporting similar results in 1995, D. K. Sokol and colleagues observed that among a group of four- to six-year-old identical twins, those who shared a chorion as fetuses were more similar to each other on 20 different personality measures than were those who developed their own chorions as fetuses.* Such effects are not limited to psychological variables such as personality and intelligence; K. Beekmans and colleagues recently noted that "biological differences are associated with type of placentation, e.g., birth weight, within-pair birth-weight differences, sex proportion, and type and frequency of congenital anomalies." Thus, the available data, while not extensive, do support the idea that experiences in utero affect the development of some traits; these findings undermine the idea that so-called identical twins are necessarily identical at birth. But how do these findings bear on the equal environments assumption?

Since fraternal twins are formed from two entirely separate fertilization events, fraternal twins *always* develop their own chorions, each twin connected to its mother via its own placenta. Thus, while a relatively large fraction of identical twins *share* a chorion, *no* fraternal twins share a chorion. As a result, fraternal twins can be expected, on average, to be less similar than identical twins, *for reasons having to do, in part, with the* environments *in which they developed as fetuses*. What this means is that behavior geneticists are on shaky ground when they conclude that differences characterizing fraternal, but not identical, twins reflect the fraternal twins' differing genes, because such conclusions require them to make the invalid assumption that different twin types develop in equally similar environments. The upshot of this situation is that whenever twin researchers compare typical identical twins with typical fraternal twins, they inadvertently generate heritability estimates that systematically exaggerate the extent to which genetic differences contribute to trait differences. Clearly, these observations call traditional interpretations of twin studies into question.

THE MYTH OF IDENTICAL TWINS

In 1810, in what is now Thailand, a rare embryological event occurred, one whose consequences shed light on the factors that contribute to trait development. Born to a Chinese father and a half-Chinese mother living in Thailand (known as Siam at the time), Chang and Eng arrived in this world joined together at the sternum by a muscular band of tissue 8 centimeters long and only half that thick. While today's surgeons would readily have separated them at birth, the requisite surgical techniques were not yet available in Chang and Eng's day. Consequently, Chang and Eng lived their entire lives completely in each other's company.

At the age of 18, Chang and Eng left Siam with a British merchant who had resolved to "exhibit" them throughout the Western world. But after three years of turning most of their earnings over to this "sponsor," the twins took charge of their own career and began charging customers for the opportunity to see them run, swim, and play badminton, all with exceptional coordination and grace. Before the 1830s ended, Chang and Eng had amassed a small fortune, which they used to buy a farm in North Carolina (where they had toured years earlier). In 1839, the twins became naturalized citizens of the United States, and they settled down with their respective wives—a pair of sisters, each of whom ultimately bore her husband nearly a dozen children!—to live out their 63-year-long lives, the longest ever lived by so-called conjoined twins. Thanks in part to the huckstering efforts of P. T. Barnum, who exhibited them in his New York City "museum" in 1842, Chang and Eng lived to become the most famous conjoined twins the world has ever known. In fact, owing to the name of their country of origin, Chang and Eng came to be recognized throughout the world as the original Siamese twins, a moniker now synonymous with the phrase "conjoined twins."

Conjoined twins develop in rare cases when embryos split in half *later* than nine days after they have been conceived. By this point in gestation, development has progressed far enough that the resulting twins share more than just a single chorion: in some cases, they also share body parts, and so are born with their bodies joined together at some spot. When conjoined twins cannot be surgically separated, they wind up living their entire lives in environments that are as nearly identical as possible. And since fraternal twins cannot be conjoined, all conjoined twins are "identical" twins, and so share identical genes as well.

What this means is that conjoined twins like Chang and Eng share identical genes *and* nearly identical environments throughout their lives. Thus, if any twins could be expected to have truly identical traits, these would be them. Nonetheless, as Stephen Gould writes, Chang and Eng "carried on independent conversations with visitors and had distinct personalities. Chang was moody and melancholy and finally took to drink; Eng was quiet, contemplative, and more cheerful." Identical traits, then, need *not* characterize even twins with identical genes who are reared in environments that are as close to identical as possible.

In fact (as I have noted), there are no such things as truly identical twins. For the same reasons that Chang and Eng had unique characteristics, genetically identical twins—or, for that matter, clones—are not photocopies of each other, either. In fact, ordinary "identical" twins can be expected to develop into adults who are even more distinct than were Chang and Eng, because unlike twins who are in each other's presence at all times, ordinary "identical" twins accumulate independent experiences throughout their lives, experiences that invariably affect their traits.

The most convincing demonstrations that twins are never identical come from well-controlled laboratory studies of simple animals raised in simple environments. The reason these studies are so convincing is related to the fact that simple organisms have fewer traits than complex organisms (think of the number of traits that characterize a worm versus the number that characterize a dog). Since simpler organisms have fewer traits on which they would have to match to be identical, two relatively simple organisms are more likely than two complex organisms to be truly identical. Therefore, if even simple, genetically identical organisms raised in identical environments *still* aren't identical, we can be confident that we will never find identical complex organisms, *even if these complex organisms are genetically identical and raised in identical environments.*

Consider, for example, a type of nematode, every normal specimen of which has exactly 1,031 cells that constitute its body. This invariant number of body cells seems, at first glance, to suggest that these worms all develop in exactly the same way. Nonetheless, *genetically identical* worms reared in *identical environments* have been found to exhibit nonidentical behaviors in those environments. Thus, despite their identical genes and environments, differences in behavioral traits emerge during the development of different worms. According to Gilbert and E. M. Jorgensen, "organisms with the same inheritance . . . and the same environment . . . still [wind up with] behavioral differences as a result of chance events during development." But how in the world can such differences arise in the face of identical genes *and* identical environments?

The famous developmental biologist Conrad Waddington coined the term "developmental noise"* to refer to the source of variations in traits that could be attributed to neither genetic factors nor to obvious environmental factors like diet, weather, or maternal health, for example. The importance of this sort of factor in the appearance of traits has been recognized by biologists since at least 1920, when Sewall Wright could not account for variations in the colors of guinea pigs' coats by considering only the effects of genetic and obvious environmental factors. Because of results like these, biologists now understand that the appearance of a given trait is not predetermined, "even when the genotype [genes] and the environment are completely specified." Thus, raising "identical" human twins in seemingly identical environments still does not ensure the development of identical traits. Instead, "noisy" factors like uncontrollable, random environmental events or variations in the timing of critical gene-environment interactions can have profoundly important effects on traits like fingerprint patterns and patterns of coloration in the eyes. In the chapters that follow, I will discuss a few of the innumerable ways and moments in which such factors can impact developing animals; given the existence and importance of such factors, it is no wonder that pairs of truly identical organisms do not exist.

In retrospect, one of the best reasons to have anticipated this result is the observation that individual animals can be asymmetrical with regard to certain traits. Even though a given animal develops in a single environment and possesses a single set of genes, still such asymmetries are common. Lewontin writes:

> The two sides of a *Drosophila* have the same genotype, and no reasonable definition of environment will allow that the left and right sides of a pupa [an immature fly] developing halfway up the side of a glass milk bottle in the laboratory are in different environments. Yet . . . the number of eye facets differ between the two sides of an individual fly.

Similarly, the next time you see a Siberian Husky with one brown eye and one blue eye—a trait that is quite common in this breed of dog but that is not *specified* in the genes—scratch your head; asymmetries like this should make us think twice about our received belief that traits reflect only genetic and environmental inputs. Such asymmetries, like musician David Bowie's one-brown-one-blue eyes,* should also encourage us to think back skeptically on the proclamations of our seventh-grade biology teachers who taught us that the color of our eyes is determined strictly by our genes.

SORTING IT ALL OUT

If even conjoined twins, with their identical genes *and* their as-identical-as-can-be environments, still don't develop identically, why do identical twins so often look and act just like each other? Why are identical twins so much *more* similar to each other than are fraternal twins?

Although "identical" twins' genes are not single-handedly responsible for the similarity of their traits, this does *not* mean that their genes do not *contribute* to the twins' similarities; they must! After all, according to Galton's conceptualization of the forces driving trait development, there are only two factors to consider: nature and nurture. And since "identical" twins share identical genes whereas fraternal twins do not, it would be extremely unusual indeed if identical twins were *not* more similar to one another than are fraternal twins; two identical sets of genes operating in a given pair of environments should be *expected* to produce more similar traits than would differing genes operating in this same pair of environments. The essence of the developmental systems perspective is that genes are essential, but not exclusive, causes of the development of traits; thus, developmental systems theorists are not the slightest bit surprised that individuals with identical genes are more similar to one another than are individuals with differing genes.

But acknowledging that identical genes contribute to identical traits does not mean that the traits are determined strictly, or even primarily, by the genes. It means only that genetic factors *influence* trait development, a contention that should be accepted by everyone working in biology or psychology these days. The truth is that development is a complex process involving genetic factors, environmental factors, *and* phenomena like developmental "noise," so acknowledging that identical genes contribute to similar traits represents only the beginning of an answer to questions about traits' origins, not a satisfying end.

Satisfying answers to questions about trait origins will explain *how* genes interact with nongenetic factors to cause trait development, and such answers can be provided only by studying *development* itself. *Twin studies, which merely "account for variation" in extremely narrow circumstances, cannot persuasively contribute to our understandings about the causes of traits.* And if we already know from the outset that genes are essential (but not exclusive) causes of trait development, then twin studies do not advance our understanding at all. Why, then, do behavior geneticists conduct them?

When I read reports of twin studies, I sometimes get the sense that their authors' goal is to demonstrate that genes are an important influence on our traits. If this *is* the goal of behavior geneticists, there is a good historical reason they adopted it: through the middle of the twentieth century, the field of psychology was enthralled by an environmental movement—

called behaviorism—that, for the most part, *did* deny the importance of genetic factors in the development of our traits. At the time, respected psychologists such as John B. Watson believed that many of our traits are nearly impervious to genetic factors. In 1930, Watson wrote the words that have since become the whipping boy of behavior genetics:

> Our conclusion, then, is that we have no real evidence of the inheritance of traits. I would feel perfectly confident in the ultimately favorable outcome of careful upbringing of a *healthy, well-formed baby* born of a long line of crooks, murderers and thieves, and prostitutes. Who has any evidence to the contrary? Many, many thousands of children yearly, born from moral households and steadfast parents become wayward, steal, become prostitutes, through one mishap or another of nurture.

In retrospect, Watson's implied inference that parents' traits and their offspring's traits are completely unrelated strains the bounds of believability; these days, scientists and the public alike understand how important genetic factors are in the development of our traits. But given the environmentalist atmosphere prevalent in America for much of the twentieth century, it is no wonder that it became important for Galton's intellectual descendants to search for real evidence of the inheritance of traits. Casual observation suggests that some traits *do* run in families, and assertions to the contrary might, perhaps, have been expected to produce the reaction they did: a mobilization of scientists to disprove the unwarranted claims.

There probably remain pockets of environmentalism that hearken back to Watson's ideology; in fact, some critiques of behavior genetics have been criticized themselves for denying (or seeming to deny) the importance of genetic factors in trait development. For the most part, though, the argument between behavior geneticists and Watsonian behaviorists is over, and the behavior geneticists won; the fact that genes influence our traits is not often questioned in scientific discourse these days. But do note that environmentalism did *not* collapse under the weight of a growing body of behavior genetics data with which it was incompatible.* Instead, as we will see, studies of trait *development* have led to the insight that all traits are influenced by genetic factors (just as they are all influenced by environmental factors). My current point is that even though genes undeniably influence trait development, new understandings of why we have the traits we do will come not from behavior genetics studies of the *differences* in traits but from studies of the *causes* of traits. There is little to be gained from studying the sources of *differences* in traits, simply for the purpose of battling the naïveté of an outmoded environmentalism.

Behavior geneticists fail to answer questions about the origins of traits partly because they try to understand the *independent* effects of genes and environments on traits when, in truth, traits arise epigenetically from the *dependent interactions* of genes and environments. Fortunately, even though "heritability" isn't what it sounds like and the twin study is *not* a fabulous tool for analyzing the impacts of genes and environments on our traits, we do know quite a bit these days about trait origins. We know what we now know from studying how traits *develop*. And the good news is that direct exploration of the efficient causes of development has produced an extraordinarily complex, but nevertheless exceedingly beautiful, preliminary picture of the epigenetic origin of traits.

Part II

BACKGROUND BASICS

Deep understanding of nature requires a close scrutiny
of the details of nature.

—John Muir

4

DEPENDENT GENES
Essential Biology and DNA

The complexities and wonder of how the inanimate chemicals that are our genetic code give rise to the imponderables of the human spirit should keep poets and philosophers inspired for millennia.

—J. Craig Venter, president of Celera Genomics, from his remarks at the White House, June 26, 2000

When I went to see the critically acclaimed film *Magnolia* by writer-director Paul Thomas Anderson, I was surprised to see in its opening sequence a story that I had read on the Internet a couple of years earlier. While Anderson had changed most of the names mentioned in the account, as well as the year in which the events were alleged to have unfolded, the basic idea remained the same. This is Anderson's version:

> The tale told at a 1961 awards dinner for the American Association of Forensic Science by Dr. Donald Harper,* president of the association, began with a simple suicide attempt—seventeen year old Sydney Barringer, in the city of Los Angeles on March 23, 1958. The coroner ruled that [Sydney's] unsuccessful suicide had suddenly become a successful homicide. . . . At the same time young Sydney [jumped off] the ledge of this nine-story building, an argument swelled three stories below. The neighbors heard, as they usually did, the arguing of the tenants, and it was not uncommon for them to threaten each other with a shotgun. . . . And when the shotgun accidentally went off, Sydney just happened to pass [outside of the apartment's window]. . . .
>
> The two tenants turned out to be Fay and Arthur Barringer, Sydney's mother and Sydney's father. When confronted with the charge . . . Fay Barringer swore that she did not know that the gun

was loaded. [But] a young boy who lived in the building, sometimes a visitor and friend to Sydney Barringer, said that he had seen, six days prior, the loading of the shotgun.

It seems that the arguing and the fighting and all of the violence was far too much for Sydney Barringer and knowing his mother and father's tendency to fight, he decided to do something. The young neighbor explained [to the police], "He said he wanted them to kill each other, that all they wanted to do was kill each other and he would help them if that's what they wanted to do."

Sydney Barringer jumps from the ninth floor rooftop. His parents argue three stories below. [Fay's] accidental shotgun blast hits Sydney in the stomach as he passes the arguing sixth floor window. He is killed instantly but continues to fall—only to find, three stories below—a safety net installed three days prior for a set of window washers that would have broken his fall and saved his life if not for the hole in his stomach. So Fay Barringer was charged with the murder of her son and Sydney Barringer noted as an accomplice in his own death. . . . Ohhhh. These strange things happen all the time.

Despite the apocryphal nature of this story, Anderson was quite right: strange things do happen all the time. And while biological activity in and around genes might not be truly "strange," such activity *does* happen all the time, and it is most definitely analogous to the events surrounding Barringer's baroque death. In both cases, complex sequences of intertwined, history-dependent events lead to an outcome whose final disposition is determined by the clockworklike unfolding of those events.

DOMINO FLOW

The idea that trait development is influenced by only two possible factors—our genes and our environments—suffices as long as no one asks the really tough questions: *How* do genetic and environmental factors influence development? What exactly happens that leads to change, and why are the changes that occur the ones that occur, instead of some other imaginable changes? No one can yet answer these questions about most traits, but it turns out that merely *asking* these questions is illuminating, as they lead to a richer understanding of the nature of the problem.

For the environment to cause a trait to develop, it must somehow affect the body; in the case of the development of a psychological trait, the environment would have to affect the structure and/or function of the brain. Similarly, for the genes to cause a trait to develop, they must *do* something that affects the body, somehow. And as soon as we begin to look at how the genes and the environment physically affect bodies, we discover a major problem with the idea that we can explain development solely with reference to these two types of factors.

The discovered problem has to do with how we should think about those "biological" factors inside our bodies—the chemicals, cells, and organs that we're made of—that are not, themselves, genes. These factors, because they are not genes per se, are not "genetic"; still, they don't initially appear to be "environmental" either. And while it might be simpler to just ignore them, the truth is that these are the factors that are always actually impacted by the genes on one side and the environment on the other; because they mediate between the genes and the environment, they are typically of central importance in trait development.

Clearly, biological levels *between* the genes and the environment are not captured by either the word "genes" or the word "environment." As a result, our language is inadequate for talking about trait development. An account of trait development that refers only to these concepts must be simplistic, since such a conceptualization cannot possibly represent the *real* complexity of nature. If the question of the relative importance of nature and nurture to trait development is cast as a question about the relative importance of genes and environments, then no wonder we have been wrestling unsuccessfully with the nature-nurture issue for so many years! A deep understanding of trait development will forever elude those who ponder this problem without considering the role of the nameless factors that are neither "genetic" nor "environmental" but that inhabit the space *between* these two. Taken together, these factors constitute what I will henceforth call the "microenvironment" of the genes and cells. This term will allow us to distinguish such factors from the factors constituting the "macroenvironment" outside our bodies.*

Complicating matters is the fact that the microenvironment contains many, many elements. As a result, a macroenvironmental event (say, seeing a wild grizzly bear) might affect a particular microenvironmental factor (say, the chemical state of your eyes), which might affect another microenvironmental factor (say, the electrical state of your brain), which might affect a third microenvironmental factor (say, the chemical state of your adrenal glands, which secrete adrenaline when you're in danger). Such events, in turn, can affect other microenvironmental factors (say, the amount of a particular hormone circulating in your blood), which might affect still other microenvironmental factors, or even genes themselves (as we will see). Sequences of events like this, in which event A causes event B, which causes event C, and so on, are characteristic of biological processes, many of which occur entirely within the microenvironments of the genes and cells. In fact, this arrangement is so common that biologists have appropriated a word from common English to refer to it: they often speak of "cascades" of events, in which each event causes the next one in the sequence, much as falling dominos push over their neighbors. Psychologists are beginning to think in this way, too. L. B. Smith writes that development can be "determined, not by some prescribed outcome . . . but as the product of a history of cascading causes in which each subsequent change depends on prior changes and constrains future changes."

How exactly should we think about the efficient *causes* of outcomes that result from cascading events? If a series of 26 equally spaced dominos are arranged in a line, and pushing over domino A (event A) ultimately leads to the toppling of domino Z (event Z), is it fair to say that event A "causes" event Z? There is a sense, of course, in which event A does cause event Z. However, to me (and to some respected philosophers who have considered this issue in great detail), it does not seem reasonable to call event A *the* cause of event Z, because many other events are involved in producing event Z as well. My reasoning, of course, is that event A is not *sufficient* to cause event Z; merely changing the orientation of *any other domino* would interfere with the cascade that ordinarily topples Z. Moreover, event A is not even *necessary* for the occurrence of event Z, because we could have toppled Z just as easily by initiating the cascade with some event other than event A (say, toppling domino J).* As a result, there seems to me to be something fundamentally wrong with asserting that event A is more the cause of event Z than are events B, J, or X. In general, whenever a cascade of events leads to a particular outcome, each and every event involved in the cascade must be understood to have played an essential role in the occurrence of the outcome, since without every single event, the outcome would not have occurred as it did.

Some philosophers prefer to call event A the cause of event Z, while labeling intervening events "background factors." These philosophers like this approach because it allows us to retain our intuitive understanding of causation. For example, this approach allows us to maintain that the cause of a murder is the murderer's behavior and not the fact that the victim was wearing nonbulletproof clothing on the day of the shooting. Nonetheless, if our goal is to prevent the murder, then the victim must be understood to have died, in part, be*cause* of his clothing, since donning a bulletproof vest would have prevented the murder. This is an unusual use of the word "cause," to be sure, but if we want to *influence* the outcome of this situation, it would be important to understand causation at this level.* Understanding a situation so thoroughly that we can effectively influence its outcome typically requires recognizing how the outcome emerges from *all* of the factors that contribute to its appearance.

Our traits are outcomes of complex cascades of events. Therefore, if our goal is to understand a trait's development so thoroughly that we can influence its appearance, we must recognize the causal importance of *all* of the factors that contribute to that appearance. In fact, genes, microenvironments, and macroenvironments each contribute in essential ways to the appearances of our traits, since altering any of these three can lead to different outcomes. In this sense, then, all three of these types of factors can be seen to *collaborate* in the production of our traits.

With that bit of foreshadowing, it is time to look at how traits actually develop. But because even a beginning understanding of trait development requires some familiarity with the workings of genes and their immediate environments, we must first learn something about molecular biology (the study of chromosomes and the genes they are made of) and cell biology (the study of cells, including the microenvironments of the genes).

LEGO WITHIN LEGO WITHIN LEGO

Bodies, as you might know, are composed of a very large number of very small units called cells. Simplifying somewhat, your brain is made of brain cells, your liver of liver cells, and your heart of heart cells. While all of these cell types are different from one another (even within a given organ there are different types of cells; for example, there are several different types of brain cells in a brain), there are some things that they all have in common. For example, cells always have a boundary made of various types of molecules (obviously, then, molecules are much smaller than cells—in fact, molecules are just collections of even smaller particles called atoms). This boundary of molecules is called a *cell membrane*; its job is to hold in all of the other contents of the cell. Among these contents is cytoplasm— the fluid in which everything else in the cell is suspended—and a nucleus, one of which floats in the cytoplasm of almost every cell. The nucleus, for our purposes, consists of another membrane—the nuclear membrane— that encloses a group of complex molecules called chromosomes.

Each of the cells in your body contains in its nucleus 46 chromosomes arranged into 23 pairs (other species have different numbers; dogs, for example, have 39 pairs of chromosomes). One of the chromosomes in each pair came originally from your mother and the other came from your father; together, these 46 chromosomes constitute your entire genetic complement, or *genome*. Importantly, although the various types of cells that make up a body are all different from one another, *each of their nuclei contains the identical chromosomes*. This general arrangement holds for living things from gorillas, whales, birds, fish, and insects to oak trees and rose bushes.

The fact that bodies are made of organs that are made of cells that contain nuclei that contain chromosomes means that there are a variety of biological factors to consider as we explore how traits develop. In particular, in addition to your body existing in a macroenvironment, your organs exist in a microenvironment containing other organs, your cells exist in a microenvironment containing other cells, and each of your chromosomes exists in a microenvironment containing other chromosomes. And, as we will see, trait development results from interactions that occur among these various components of the complex system that is your body.

THE TWISTED LADDER

Understanding the roles that chromosomes play in trait development requires an understanding of what chromosomes are and of how they work (although oddly, many people seem comfortable labeling traits "genetic" without even a rudimentary understanding of how the "genes" that make up chromosomes work!). Chromosomes are complex molecules consisting (predominantly) of DNA, which is made of two long chemical strands that are *weakly* bonded to each other. Each strand consists of a sequence of "bases" that are *strongly* bonded together in a long chain (these bases, also called "nucleotide bases," are molecules that can be thought of as having particular shapes; more detail than this is not necessary for our purposes). We can schematically picture DNA as a dangling (vertical) string of magnetic beads, made up of only four types of beads; in this case, the bases are analogous to the beads, and the strong bonds are analogous to the string running through their centers that holds them together. The weak bonds, then, can be imagined as the horizontal, magnetic bonds that keep beads on one strand together with corresponding beads on the other strand; these weak bonds between individual bases keep the two strands of DNA weakly bonded to each other. DNA contains only four different kinds of bases, which are typically known as A, C, G, and T (these are the initials of their real names—adenine, cytosine, guanine, and thymine).

What makes DNA molecules special is that they are able to replicate themselves, which is to say that they can produce exact copies of themselves. This is an absolutely crucial feature of DNA, because when a cell undergoes division, it produces two cells where there was previously only one. And if every cell in a given body must have the same complete set of chromosomes, then just before a new cell is formed, a new set of identical chromosomes—which will ultimately reside in the new cell's nucleus—must be created. So, how did Mother Nature solve the problem of getting a complex molecule like DNA to make perfect copies of itself?

The process of evolution has provided us with bases that can form *weak* bonds with only specific other bases located on another DNA strand. The "shape" of base A allows it to form weak bonds with base T, and never with bases C or G (although any two bases can be *strongly* bonded with each other in a single strand). Similarly, C and G are complementary: they can form weak bonds with each other. To see how such an arrangement solves the problem of replication, imagine taking a hypothetical, short piece of DNA and breaking apart the weak bonds holding its two strands together. If, in peeling apart the strands, we find that one of them (say, strand 1) is made up of bases G, A, and C strongly bonded to one another in sequence (G–A–C), then strand 2 must be made up of bases C, T, and G strongly bonded to one another in sequence (C–T–G). It has to be this way, or the two strands composing this piece of DNA would never have "fit" together in the first place.

Now we can see how this arrangement solves the problem of replication. If we set the G–A–C strand loose in a soup of free-floating bases, eventually (or quickly, if the proper facilitating enzymes are present) a free C will link up (weakly) with the G on this strand, while a free T will link up (weakly) with the A. These once-free bases—the C and the T—are now neighbors, and as soon as they are close to each other, a strong bond will form between them; this is the start of a new strand. In turn, a free G will link up (weakly) with the final C in our original strand, and this G will then form a strong bond with the T that is now its new neighbor. At this point, the newly formed C–T–G strand, which is a perfect complement of strand 1 and a perfect replica of strand 2, can break away from the original strand (to which it was only weakly linked anyway) and float off on its own. If the new strand now floats loose in a soup of available bases, of course, the same sorts of processes ensure that the next strand produced will be an exact copy of the original strand, G–A–C. A beautiful solution to the problem of replication, is it not?

THE MEAT OF THE ARGUMENT: PROTEIN PRODUCTION

DNA does only one thing: it provides the information* needed to produce proteins. Proteins are molecules that perform an astoundingly wide variety of important functions in our bodies; we'll discuss some of these functions later. Like a strand of DNA, a protein consists of a series of components linked together in a long chain. In the case of proteins, though, these components are called amino acids, not nucleotide bases (in addition, proteins are not double-stranded and they cannot replicate themselves). The nature of the DNA "code" is such that a sequence of three (strongly bonded) bases along a single DNA strand almost always specifies exactly one particular amino acid (for example, the three-base sequence G–A–C refers to the amino acid leucine). Given that DNA is composed of four bases, there are 64 distinct three-base combinations that can be generated. How is this "information" used to produce proteins?

First, DNA produces a single strand of RNA through a process exactly like that just described for DNA replication. (For our purposes, RNA can be thought of as being just like DNA except that it always comes as a single strand, never as a DNA-style double strand*). This RNA strand then migrates out of the nucleus and into the cell's cytoplasm, ultimately floating over to one of the special structures in the cell that actually manufactures proteins. A large number of these structures, called ribosomes (pronounced RYE-bo-somes), float around in a typical cell's cytoplasm, participating in the *translation* of RNA and actually constructing the proteins in living bodies. Thus, RNA is "read" by the cellular machinery at the ribosome: each time three RNA bases have been read, the amino acid that corresponds to these three bases is fitted on to the end of the new protein being constructed. In this way, the long chain of bases constituting an RNA molecule is used to produce a long chain of amino acids—a protein. *This is what chromosomes do, no more, no less.*

As it happens, each strand of RNA that migrates out of the nucleus contains a string of bases that is much longer than that needed to construct a single protein. Given this fact, how does the cellular machinery know when it has finished constructing a complete protein? It turns out that there are three three-base codes (A–T–T, A–T–C, and A–C–T) that say to this machinery "END PROTEIN CHAIN HERE" and one more that says "START BUILDING A NEW PROTEIN CHAIN HERE." It is particularly important for such codes to exist, because codes for a new protein chain do not necessarily begin just after the end of the codes specifying the last chain. Instead, on a single chromosome, there is almost always a bunch of gobbledygook between the end of one set of meaningful codes—called a "cistron"—and the beginning of the next set of meaningful codes (the next cistron). In fact, many biologists currently believe that the vast majority of our DNA codes for nothing at all. "In humans, as little as ten percent of the DNA is translated"; our 80,000 cistrons are "scattered throughout the genome like stars in the galaxy, with genomic light-years of noncoding DNA in between." Clearly, a mechanism must exist whereby useful segments of DNA can be identified as such and distinguished from gobbledygook noncoding segments; the "START" and "STOP" codes provide such a mechanism.

There is something very special about proteins, something that makes them worthy of their central position in the chain of events lying between DNA and traits: they can have unique shapes. In fact, proteins perform the jobs they do in our bodies *because* of their unique shapes. And since a major determinant of a protein's shape is the order of its amino acids—an order specified by the order of bases on a DNA strand—it might seem at first glance that a protein's shape is determined exclusively by genetic factors.*

However, a protein's shape is *not* determined exclusively by the ordering of its amino acids. As Timothy D. Johnston notes, the shape of protein molecules *also* "depends on aspects of the intracellular environment such as temperature and pH [the temperature and the acidity inside the cell]." Thus, a particular segment of DNA merely contains amino acid sequencing information that, *when used in a particular environment*, specifies a protein that can do a specific job (because of its unique shape). After a collection of amino acids has been strung together in a particular order, all further development is influenced by nongenetic factors.* Thus, DNA cannot be thought of as single-handedly producing complete, functional proteins; it certainly cannot be thought of as producing full-blown traits. Johnston reports the reality concisely: "Between amino acid sequences and behavior is a long and tortuous developmental route, most of whose details are at present unknown."*

An example might be of use here. One of our biological traits commonly thought to be "genetic" is hair color; this assumption probably results from the observation that hair color runs in families and from the fact that, at first glance, there are no salient environmental factors that obviously influence it. But because genetic factors can do no more than specify amino acid sequences—and because hair colors are not themselves amino acid sequences—such factors cannot single-handedly determine hair colors. Thus, if we don't think environmental factors can influence hair color, it must simply be because we haven't yet become aware of their effects.

Hair color—like eye and skin color—is determined by the presence of melanin. Melanin is not a protein;* instead, it is formed as an end product during the normal biological breakdown of a particular amino acid called "tyrosine." What this means is that environmental factors that affect the breakdown of tyrosine also affect coloration. In the case of hair, the degree of natural melanin accumulation turns out to depend, in part, on the relative concentrations of copper in the cells that are producing the hair; dark hairs contain higher amounts of copper than do light hairs. Thus, nongenetic factors such as diet affect hair color: should the intake of copper fall substantially below a fraction of a milligram per day, new hair emerges successively less dark. Restoring sufficient copper to the diet reverses this trend (common foods that contain relatively large amounts of copper include chocolate, mushrooms, shellfish, and nuts, among others).

The effect of copper intake on hair color underscores two points. First, hair color is not determined strictly by genetic factors, because such factors can only determine amino acid sequences and hair color—like every other trait—is not itself such a sequence. Instead, nongenetic factors play important roles in determining our hair colors as well. Second, the effect of the environment is not obvious in this case, either because most of us are completely unaware of our dietary intake of copper, or because *variation* in copper intake is normally so small that we cannot see the effects of such variation.* Regardless, diet *is* an environmental influence on hair color, even if circumstances render this influence relatively hard to see.

Some headway has been made in discovering the details of the developmental routes that run between genetic factors and some traits; still, for many traits the details remain unknown. Nonetheless, we should not assume that lack of information about *how* the environment exerts its effects means that there *are* no such effects. For example, even in the absence of detailed information about how it works, we know that hair color in Himalayan rabbits, like some other mammals, can be influenced by temperature. Normally, the fur of these rabbits is white except at the rabbit's extremities—its feet, tail, ears, and nose—where its fur is black. If, however, the white hairs on this rabbit's back are plucked out and an ice pack subsequently placed on the bare spot while the hairs regrow, the new fur that appears on the rabbit's back will be black.* Thus, even though we often think of mammalian hair colors as being genetically determined, this trait can clearly be influenced by macroenvironmental factors such as diet and temperature, regardless of whether or not we understand the details of how this happens.

Biologists no longer question the following two facts: (1) a bit of chromosome can do no more than provide information about the order of amino acids in a chain; and (2) traits are *constructed* in cascades of steps—many involving nongenetic factors—that lie between amino acid sequencing and final trait production. Given these facts, one conclusion is inescapable: genetic factors cannot themselves cause traits, even traits widely thought to be "genetic," such as hair color, eye color, or body type. But if genetic factors can never produce full-blown traits independently of environmental factors, it would seem that we have already reached the end of our story. And in a sense, we have: just as no single domino in a series can be called *the* cause—or even the most important cause—of an outcome, neither can a genetic factor alone be *the* cause—or even the most important cause—of the development of a trait. Instead, genetic and nongenetic factors determine traits' appearances *collaboratively*.

So why is this book not ending here? For a couple of reasons. First, even armed with an understanding that chromosomes merely provide amino acid sequencing information and that traits are produced by cascades of events, one might still conclude that genetic factors are singularly important, simply because their contributions to cascades precede the contributions of nongenetic factors. As it happens, this is not the case at all, but understanding as much requires knowledge of some recently published, fascinating research results. Second, learning that genes do not cause traits in simple or direct ways might leave an unseemly hole in the middle of most people's conceptions of where traits do come from. And as nature abhors a vacuum, leaving this hole unfilled risks producing an uncomfortable sense of ignorance that some readers might combat by reverting to old ideas: in the absence of a *better* understanding of how traits arise, the notion that genes cause traits might seem to be better than no understanding at all. Fortunately, some things are now known about how nongenetic and genetic factors act together—co-act—during development to generate our traits. And since exposure to examples of such co-actions will help drive home how *essentially interdependent* genetic and nongenetic factors are on one another, on we go. But before beginning the discussion of development that will lead us through these examples, a brief consideration of the word "gene" is warranted.

WHAT IS A GENE?

You will have noticed a striking absence in the foregoing "primer" on genes: the word "gene" doesn't appear in it until the last paragraph! You might be surprised to learn that, in part, the reason for this omission is that *there is no agreed-upon definition of the word "gene."* The way the media barrages us with information on "genes," one could certainly be excused for believing that the word refers to a definite thing. This, however, is not the case.

In fact, the first person to use the word "gene" was the Danish biologist Wilhelm Johannsen in 1909. For Johannsen, "gene" was a generic term for the heritable "factors" that Mendel had postulated in 1865 to explain the results of his experiments on trait inheritance in pea plants. At the time, everyone understood that the "gene" was a hypothetical construct. *Something* provided baby organisms with inherited developmental information, and even though no one knew what this something was (or even if it was a material substance), it was agreed that whatever it was, it would be called a gene. By the end of the 1920s, it was clear from the studies of Morgan and others that these "genes" were chemical structures that resided in the nuclei of cells and that were related, somehow, to chromosomes. By the mid-1940s, some scientists understood that the genes were pieces of DNA, though no one could yet comprehend how DNA could possibly carry all the information it would need to carry to serve as the genetic material. Then, in 1953, James Watson, Francis Crick, and Maurice Wilkins discovered the dual-stranded, twisted structure of DNA, for which they won the 1962 Nobel Prize in Medicine or Physiology. This discovery provided an initial understanding of how DNA was capable of being the elusive genetic material. So we know today that DNA is the stuff of which "genes" are made. But the question remains: How *much* DNA makes up a gene? How many triplets of bases constitute a single gene?

In considering this question, Richard Dawkins defined a gene as "any portion of chromosomal material . . . which is small enough to last for a large number of generations." This was a reasonable definition, given his goal of understanding the role of genes in evolution.* However, other definitions appear equally reasonable, given other goals, and none have proven satisfactory to all those thinking about this problem. It might seem at first glance that we should define the word "gene" as we define the word "cistron": a "length of DNA encoding a specific and functional product, usually a protein." In fact, many scientists think of genes in just this way (when I use the word "gene" in the rest of this book, I will almost always be referring to a cistron). Nonetheless, even this definition is problematic. We can see why "cistron" and "gene" cannot refer to the same thing by considering the recent—and amazing—finding that some cistrons encode *more* than one functional product.

To understand how this is possible, we must consider the internal structure of a cistron, which, it turns out, is rather complex. In living things other than bacteria and algae, cistrons contain "meaningful" sequences of bases that code for amino acid orders; such meaningful sequences are called "exons." But dispersed among the exons in a single cistron are additional base sequences called "introns," sequences that most biologists believe do not code for anything at all.*

Imagine that batter for your favorite chocolate-chip cookies is made by mixing two cups of flour, one tablespoon of chopped peanuts, one cup of butter, one cup of chocolate chips, one cup of sugar, and two eggs. The way "information" is contained in DNA is exactly like the way information for making your cookie batter is contained in the following recipe:

> Mix two cups of flour, one cross related two cups tablespoon of chopped bag element peanuts, one cup of case honest butter, one cup of penguin green chocolate flint chips, one walking spoon cup of nail bank sugar, and two canvas eggs.

I find it almost unbelievable that nature does this in this way, but nature doesn't care what I think: scattered almost universally among our exons are seemingly meaningless introns. (There are, on average, about 10 introns within each cistron, and as much as 90 percent of a given cistron can consist of introns.) Obviously, then, something must be done to ensure that the "junk" non-information contained in introns is edited out and discarded before protein production begins; only information in exons should be retained, decoded, and put to use in protein construction. Nature has addressed this requirement with a phenomenon called RNA splicing, a wondrous collection of processes, some of which are now understood to *control* or regulate what "genes" do.

In RNA splicing, the non-information encoded in introns is cut out by temporary structures—called "spliceosomes" (pronounced splice-oh-somes)—that are formed for this purpose within a cell's nucleus. When a cistron is to be decoded, a piece of RNA that complements the *entire* cistron is produced, complete with portions that complement both its introns and its exons. This RNA—actually pre-RNA, because it still contains gibberish thrown in by the introns—then migrates over to a spliceosome that systematically cuts the introns out of the chain and joins ("splices") the exons together, ultimately producing a piece of "mature" RNA made up exclusively of uninterrupted exons. (It is this "mature" RNA that then migrates out of the nucleus and over to a ribosome, where its "information" is decoded and used to construct a protein.) In our recipe metaphor, this would be like cutting "case honest" out of "one cup of case honest butter," and splicing "butter" to "one cup of" in order to yield the "mature" instruction, "one cup of butter."

As if this arrangement is not extraordinary enough, recent research has revealed something even more astounding. In a process called "alternative splicing," the spliceosomes in different cells can do different things with the *same* pre-RNA, thereby generating two or more different proteins from the code of a single cistron. Thus, spliceosomes can edit a piece of pre-RNA in one cell type so as to produce a particular strand of mature RNA, and they can edit this *same* pre-RNA in a different cell type so as to produce an entirely different strand of mature RNA! Upon being transported out of the nucleus and then to the ribosomes, these mature RNA strands would contribute to the production of distinctly different protein forms.

What this means is that the same block of "information" can be used to create *two (or more) different products*. Thus, our unprocessed, "immature" cookie recipe, above, could be spliced in some contexts to generate the recipe for your favorite chocolate-chip cookies, but the *same* "immature" recipe could also be spliced in a different context to generate the following entirely different "mature" recipe for peanut butter cookies:

> Mix two cups of flour, two cups of peanut butter, one cup of sugar, and two eggs.

Amazingly, cistrons on DNA, too, code for immature RNA that can be spliced in different ways in different contexts, giving rise to different types of mature RNA and ultimately, to different protein products.

S. G. Amara and colleagues have provided an excellent example of such an arrangement. These researchers discovered that *the same pre-RNA* can be spliced to code either for an amino acid chain involved in the body's regulation of calcium or for a neurohormone (it doesn't matter for our purposes how amino acid chains and neurohormones differ—what matters is that they do distinctively different things). Subsequent research revealed that which molecule is produced *depends on the type of cell that is doing the producing.* In particular, when the pre-RNA is spliced in cells that make up the nervous system, the neurohormone is produced; when the identical pre-RNA is spliced in other cells, the calcium-regulating amino acid chain is produced. Thus, the product built using a given length of DNA *depends* on the tissue doing the building; that is, *different outcomes can be expected when a single cistron is decoded in different contexts.* And since specific spliceosomes effectively make different "decisions" about what product to build, the resulting outcome "does not . . . inhere in the DNA code, but emerges from the interaction of many parts of the cellular machinery." As incredible as this arrangement might be, it is now known that this sort of alternative RNA processing is actually quite common, occurring during the "reading" of as much as one third of our DNA.

The existence of alternative splicing (and other, equally amazing processes of RNA manipulation*) effectively means that the words "cistron" and "gene" cannot be used interchangeably. These processes ensure that the RNA produced from a cistron can be edited to contribute to the construction of many different gene products, sometimes numbering into the hundreds; this implies that "genes" for different products can be embedded—sometimes *overlapping!*—in the same cistron. Thus, genes and cistrons cannot be one and the same thing. So, we are left with our original query: What is a gene?

In considering this question, E. M. Neumann-Held writes:

> The analysis of the molecular mechanisms of [protein] expression shows quite clearly that there is no fundamental way by which the classical . . . gene concept could be applied to DNA segments. One focuses at the same bit of DNA, and different structures and functions can appear. One focuses on different levels of the expression process . . . and again different structures and functions appear. Introns can become exons, which can become promoters, and so on [promoters are sections of DNA that, when engaged by specific molecules, cause other sections of DNA to be decoded]. . . . There is no general rule that a particular sequence [of DNA] codes for only one [gene product].

Reaching the same conclusion, E. F. Keller writes, "[T]o an increasingly large number of workers at the forefront of contemporary research, it seems evident that . . . the sheer weight of the findings . . . [has] brought the concept of the gene to the verge of collapse. What is a gene today? As we listen to the ways in which the term is now used by working biologists, we find that the gene has become many things—no longer a single entity." R. Gray concurs, pointing out that "a gene can only be functionally defined in a specific developmental context."

Genes, as we usually imagine them, rarely exist in any sort of a coherent state in the DNA, waiting to be decoded by automatonlike cellular machinery. Instead, small pieces of DNA—pieces that are not "genes" themselves, because taken individually, they do not code for complete, functional molecules—are typically mixed, matched, and linked to produce various temporary *edited* RNAs. Of most importance, the cellular machinery responsible for this editing is "sensitive" to its context. Thus, this machinery effectively "interprets" unedited, ambiguous cistrons, thereby affecting how genetic information is used. Such contextual dependence renders untenable the simplistic belief that there are coherent, long-lived entities called "genes" that *dictate* instructions to cellular machinery that merely constructs the body accordingly. The common belief that genes contain context-independent "information"—and so are analogous to "blueprints" or "recipes"—is simply false. The existence of alternative splicing ultimately requires us both to change how we think about what "genes" are and to broaden our understanding of the ways in which cells can be influenced by interactions with their local environments. For all the facile use of the word "gene" in our day-to-day conversations and news reports, it remains impossible to define to everyone's satisfaction exactly what a gene is. As Michel and Moore note, "even today the gene is still, in part, a hypothetical construct."

5

RUNNING NOSES, SMELLING FEET
A Primer on Embryology

My mother was relatively tolerant of my junior-naturalist antics as a child, but she had her limits. When a second-grade field trip to a nearby lake netted me a few inch-long tadpoles, Mom was willing at first to let my sister and me throw them in a bathtub of water to see if they would turn into frogs in captivity. But as the days wore on and my sister and I found only slightly larger tadpoles in our bathtub each morning, Mom became increasingly disconcerted about having an aquarium in lieu of a bathtub. When she had finally reached the end of her rope, she asked my father what he thought they should do.

The next evening, Dad returned home from his medical practice with a dose of what I later learned was a thyroid hormone, thyroxin. My sister and I watched him empty a capsule into our bathtub and then were hurried off to bed. And the next morning, paddling around in our bathtub, were three newly metamorphosed frogs, the tails of their abbreviated youth still receding.* Magic. It is very difficult to ever know for sure if—and if so, how—one's specific early experiences contribute to the development of one's adult traits. But sometimes I can't help but imagine that my current interest in development is connected with the magic I witnessed in my bathtub when I was seven.

These days, many people have a general understanding of how animals develop. First, a sperm penetrates an egg to produce a zygote. Development then proceeds by a combination of cell division, which ultimately leads to growth and differentiation, which ultimately leads to specialization (that is, different body parts having different functions). Together, cell division and differentiation give rise to the structure of our bodies and to the internal structure of each of our body parts. As a result, animals are born with bodies and brains that are typical of their species. Many people believe that it is only after birth that environmental stimuli begin to impact development, so that nurture now joins nature in sculpting the animal as it

82

moves from infancy to adulthood. As usual, though, a closer look reveals a significantly more complicated picture each step of the way.

So how does development *really* unfold? Armed with a grounding in some basic biology, we are now ready to take a closer look at how genes, microenvironments, and macroenvironments interact during development to produce our traits. Since our characteristic features begin to develop at fertilization, a brief initial look at fertilization itself is in order, if only to convey how much more complicated the process is than most of us typically realize. This will be a common theme as we look more deeply into biological development: the processes driving development and producing traits are breathtakingly complex.

EMBRYOLOGICAL ESSENTIALS

Think about fertilization as most of us are typically taught about it. Our basic education in these matters leaves us imagining that fertilization is a process that has more to do with physics than biology: the sperm with the greatest momentum when it encounters the egg, penetrates it, like a BB entering a marshmallow, right? Relegating the technicalities to parenthetical statements that you can skip if you're not interested in how incredibly complicated this process *really* is, here's the simplified essence of the biochemistry of fertilization (in sea urchins). A chemical (resact, a small 14-amino acid peptide) in the "jelly" that envelops the egg is just the right shape to be able to stimulate a receptor in the sperm's cell membrane. The stimulation of the sperm's receptor produces a cascade of chemical events in the sperm: the receptor stimulation activates a molecule (a G-protein), which in turn activates a second molecule (phospholipase C), which splits a third molecule into two smaller molecules. One of the two resulting molecules (diacylglycerol) decreases the acidity of the sperm (by activating a proton pump), making it move faster (by increasing its rate of metabolism). These events must occur for fertilization to be successful. Later on in the process, "the same biochemical pathway used by the egg to activate the sperm is used by the sperm to activate the egg." But in the latter case, increased metabolism in the egg is accompanied by the initiation of DNA and protein synthesis, events that signify the true beginning of development.

Is there some point to this story besides the fact that biological processes are marvelously—if outrageously—complex? Yes. Gilbert points out that "fertilization is the archetypal interaction between two cells." As a result, we can expect interactions between any two cells to have features in common with the fertilization scenario just depicted. And for our purposes, the most important feature of this scenario is that the events that unfold during fertilization do not arise solely *within* any one cell (as they would have to if they were being "directed" by genetic factors alone). Rather, they are the consequence of nontrivial *interactions* between the sperm and the egg. Many developmental scientists now believe that such interactions characterize normal development *in general*. The idea, then, is that the impetus to develop—or to develop in a specific way—arises not *within* a given cell (for example, from the genes), or from any single stimulus in the environment, but instead from the interactions *between* two or more components that constitute an integrated *system*. Thus, development depends on interactions that occur when the components of a system are in proximity to one another at the right time.

After your dad's sperm fertilized your mom's egg, the zygote that you were replicated itself, dividing into two cells, each of which is said to be a "daughter" of the original cell. These daughter cells then each divided, resulting in four daughter cells, which soon became eight cells, then sixteen cells, and so on, ultimately developing into the community of a thousand billion cells that you are now. But somewhere along the line, you had to change from a small clump of identical cells into a much larger, integrated collection of about 350 discrete cell types—including skin, heart, and brain cells—*all of which are distinctly different from one another.** This process is called "differentiation," and while scientists can now trace normal embryological development in detail, we remain largely ignorant of the mechanism by which differentiation occurs. One thing we know for sure, though, is that a cell's differentiation is *dependent* on factors arising in other cells in its environment; the developmental information that dictates how a cell will differentiate does not reside inside the cell, and so it cannot be provided by the cell's genes. But to understand how we can be sure of this, you'll need a slightly more detailed understanding of embryological development.

Because development in invertebrates is a simplified version of development in vertebrates, I can impart the needed understanding by describing the development of a simple invertebrate—the sea urchin. Keep in mind, though, that the key features of embryological development are the same in all animals, from sea urchins to humans. Thus, the seemingly weird jargon I'll be introducing—"blastula," "ectoderm," "induction"— isn't from a bad science fiction movie but rather describes real stages you went through during your embryological development, real parts of you when you were an embryo, or real processes that affected you as you grew.

After about 10 divisions, the approximately 1,000 cells that make up typical sea urchin embryos arrange themselves into a hollow ball that is one cell thick all around; at this point, all of the cells are still identical.

This structure is called a *blastula*. As the cells of the blastula continue to divide, they begin a process called *gastrulation*, which culminates in the formation of a *gastrula*, an embryo that still contains a hollow center, but that now consists of three layers of cells surrounding that hollow. At some point in their development, *all* animals are in the form of a three-layered gastrula, but the process by which this occurs varies across species. Still, gastrulation always involves the movement of cells from location to location in the embryo, as the organism rearranges itself. But if cells are moving around during development, how can we ever trace what happens to them as development unfolds?

It is possible to inject a dye into individual cells (or groups of cells) in embryos and then to examine the organism at a later stage of development; this allows us to see what becomes of the dyed cells as the animal matures. It turns out that in normal embryos, lots of development *appears* to be predetermined, insofar as certain cells always end up in certain locations in the body, serving certain functions. For instance, using the dyeing technique, it has been shown that under normal circumstances, cells in the most interior layer of the human gastrula eventually become cells that line the inside of the throat, stomach, and intestines (among other things). Cells in the most exterior layer of the gastrula normally become hair, nails, skin, and brain. And, as you might have guessed—since we've just seen that the most internal aspects of our bodies develop from the innermost layer of the gastrula and that the most external aspects of our bodies develop from the outermost layer of the gastrula—cells in the middle layer of the gastrula normally become the cells of the skeleton, muscles, and blood. Thus, the dyeing technique has allowed biologists to draw what they call "fate maps," diagrams of embryos that indicate which parts of the embryo normally give rise to which parts of the adult body.

But note the tension between the meaning of the word "fate"—"that which is inevitably predetermined"—and my extensive use of variants of the word "normal" in the preceding paragraph. In the words of embryologist Lewis Wolpert:

> The fate map should be regarded rather like a train timetable—it tells you only what will normally happen. It in no way means that other outcomes are not possible, particularly if the system is perturbed by say, bad weather, or a strike, or in the case of embryos, by experimental manipulations.

Thus, "fate map" is an unfortunate name, since there is nothing inevitable at all about the final form and function of particular cells. Instead, that form and function is determined by the complete set of circumstances along the developmental pathway that leads to the differentiation of the cell. But given that the final form of a cell is not, somehow, predetermined, what factors contribute to the determination of that form?

INSIGHTS FROM THE NEWT MONSTER: THE DISCOVERY OF INDUCTION

In 1935, the German embryologist Hans Spemann won a Nobel Prize (the only one ever awarded to an embryologist) for work that he had done 11 years earlier on newt embryos. In 1924, it was known that removing some of the ectoderm—the outer layer—from an amphibian gastrula would cause the gastrula to develop abnormally. Spemann's first significant study in this series involved removing some of the ectoderm from a gastrula and watching to see if *it* could develop normally in the absence of the rest of the gastrula; it couldn't. Thus, Spemann was left with the preliminary hypothesis that normal development of the ectoderm requires the presence of the rest of the gastrula, and that normal development of the rest of the gastrula requires the presence of the ectoderm. To explore this hypothesis further, Spemann conducted some truly ingenious experiments that revealed that even after a gastrula is formed, the fates of the gastrula's cells *still* haven't been sealed.

Spemann first cut a flap of ectoderm from a developing newt gastrula, surgically removed some of the underlying cells of the mesoderm—the middle layer—and then placed the ectodermal flap back into its original position. After letting this embryo develop for a while, Spemann began to notice abnormalities in the area he had disturbed. Such a finding might not be surprising; one might expect such harsh treatment to damage a fragile organism and to preclude normal development. So, Spemann redid the experiment with a minor change. This time, he removed a flap of ectoderm (as before), scooped out the underlying mesoderm (as before), then tamped the scooped-out mesoderm back into its original position, and finally replaced the ectodermal flap. Remarkably, despite this intrusive procedure, a normal newt still developed. This suggested to Spemann that his previous results were not caused by unacceptably harsh treatment. Instead, he concluded, his first thought could still be correct: normal development of the ectoderm might rely, somehow, on the presence of the mesoderm. But before a first-class scientist like Spemann would assume that his hypothesis was correct, he had to conduct the experimental pièce de résistance: a transplantation experiment.

In his final experiment of this series, Spemann took two newt gastrulas and prepared one of them ("the host") to receive a transplant from the other ("the donor"). First, from the donor embryo, he cut away a flap of ectoderm that he knew would normally develop into brain tissue; then he scooped out the mesoderm under this flap. Next, from the host embryo, he cut away a flap of ectoderm that he knew would *not* normally differentiate into brain tissue, and he scooped out some of the mesoderm under *this* flap. Finally, he transplanted the donor's mesoderm into the space created by the removal of the host's mesoderm, subsequently replacing the host's ectodermal flap. Spemann then let the host embryo develop for a while. When he looked at it later, he found an embryo developing with *two* brains! Not only that: the host embryo ultimately developed a complete second head! What in the world was going on here? Obviously, one of the heads had resulted from normal development, but what had produced the second head?

As Spemann—and everyone who came after him—saw it, there was only one way to understand these results. Somehow, the transplanted mesoderm caused the overlying ectoderm to differentiate into brain tissue; Spemann called this process "induction." This finding means that even though, under ordinary circumstances, a particular portion of an embryo's ectoderm is "fated" to develop into brain tissue, *any* portion of ectoderm is capable of differentiating into brain tissue if it has the proper mesoderm underneath it. This, in turn, means that the final form of an ectodermal cell (or as it turns out, any cell) is not determined solely by information contained *within* the cell. Thus, the normal "fate" of an ectodermal cell is neither inevitable, nor genetically determined. Instead, differentiation results from interactions between neighboring cells; the final form of an ectodermal cell is profoundly influenced by some sort of signal received from the neighboring mesodermal cells in its microenvironment. In his autobiography, Spemann put it this way: "We are standing and walking with parts of our body which we could have used for thinking if they had been developed in another position in the embryo."

Since Spemann's work, countless embryological experiments have demonstrated how differentiation depends on interactions among neighboring cells; we now know that cell types are *induced* in embryos by factors arising in neighboring cells. An example is the induction of the lenses in your eyes. Early in development, two protrusions of brain tissue inside the embryo—one for each eye—begin to grow outward toward a layer of ectoderm on the surface of the embryo (most of this ectoderm will ultimately differentiate into the cells of the head). When these brain tissues grow enough to actually touch the ectoderm, the ectoderm at those two sites of contact responds by growing in toward the brain tissues. In turn, the brain tissues then retreat (to give the ectoderm room to grow in), forming two cup-shaped indentations around the advancing ectodermal growths. A blob of ectoderm is then pinched off from the end of each of the growths; at this point, each ectodermal blob lies in its own "eye cup" made of brain tissue. This cup-shaped brain tissue (which ultimately becomes the retina, the light-sensitive tissue at the back of each eye), *induces* its cupped ecto-derm to become the eye's lens. We know this because transplanting addi-tional brain-derived eyecups into the vicinity of ectodermal tissues—even ectodermal tissues that ordinarily would not become lenses—causes those tissues to differentiate into additional lenses.

Spemann's work generated decades of embryological research designed to try to understand *how* different tissues interact with one another. And while it is now clear that differentiation results from interactions between neighboring cells, it has not yet been possible to elucidate the exact mech-anism by which some cells induce differentiation in other cells. Thus, differentiation remains one of the most important unsolved puzzles in developmental biology to this day. Nonetheless, despite these gaps in our understanding, Spemann's work on induction (and related work by his fol-lowers) led to an uncontroversial principle that will help us address the riddle of differentiation in the future: the structure and function of a cell is influenced not only by factors inside the cell but also by the interactions that the cell has with its neighbors in its microenvironment.*

As fascinating as Spemann's work was, some readers might object that weirdo manipulations of development—manipulations that, in Spemann's case, gave rise to two-headed monsters—cannot illuminate *normal* devel-opment, because normal development does not entail Spemannesque transplantation of various tissue types. Thus, reactions to the story of cell fates often include the following question and conclusion: If, in normal development, certain cells *always* induce certain other cells to differentiate in a certain way, why bother insisting that a cell's fate isn't inevitable? For all intents and purposes, in normal situations, a cell's fate *is* determined. Similarly, one might argue that if normal variation in the amount of cop-per we eat is extremely small, we can ignore the effect of this factor on hair color, because the variation we see in people's hair color must, therefore, not be accounted for by dietary variations. But a closer look can reveal the value of studying abnormal development.

In fact, medical breakthroughs are more likely if one carries around in one's head as complete an understanding as possible of how biological sys-tems really work, as opposed to a more limited understanding of how they work in typical situations. Many people—myself included—believe it is good to intervene in natural events if the effects of those events are disas-trous for human beings (that is, it was a good thing to develop a vaccine against polio). These people are interested in trying to eliminate scourges on mankind by manipulating natural events (including such things as embryonic development). Were we to ignore knowledge about development in abnormal or manipulated situations, we could easily miss out on opportunities to intervene in development in ways that might improve the human condition. And such a perspective is not important only for people in a position to make medical breakthroughs; insisting that it is enough to understand "normal" development blinds us all to possibilities that could improve our lives. It might be that certain kinds of "abnormal" experiences are just the sorts of experiences we need to achieve our goals. We would never discover the positive impact of such experiences if we insisted on studying *only* "normal" development.

In a recent article in *Developmental Science*, Linda B. Smith offered a concise recapitulation of what developmental psychologists should take away from studies of embryological development. She wrote:

> Development is *not* a matter of just growing, or tuning, or refining, or filling in a blueprint. Rather, real causes operating in real time literally *make* fingers and toes out of processes and stuff that are not fingers and toes. . . . [The undifferentiated cells of a gastrula] are *not* marked by their internal genetic structure; they cannot be since all the cells are the same. Instead, they are marked by their position in the mass . . . the processes that make some cells fingers and some cells toes do not start with the genes. . . . The lesson from embryological development is this. New forms are created in a *history* of events, in the *contexts* they create, and out of the *general* processes of life itself.

Thus, it is out of the interactions of genetic and nongenetic factors that our bodily traits begin to emerge while we are embryos. But even as our physical characteristics continue to develop by way of such interactions, it is the development of a particular organ—the brain—that ultimately gives rise to our psychological characteristics. *All* of the psychological traits that characterize infants, children, or adults—including species-typical traits like the ability to see in color, or unique individual traits like the ability to gracefully handle awkward social situations—are influenced by the structures and chemistries of our brains. As a result, an examination of how brains come to be as they are will be worthwhile. The story of brain development is fascinating in its own right, but it is also a story that will help me introduce some of the nongenetic factors that contribute to the appearance of our psychological traits.

6

BRAIN BASICS
Some Revelations
of Developmental Neuroscience

At the tender age of 18, a French Canadian man named Sylvère entered military service. World War II was raging overseas, and he was not about to let the fact that he had begun to have occasional and very brief lapses in consciousness stop him from contributing to the war effort. Unfortunately, while the world finally saw the liberation of Europe, Sylvère's epileptic attacks continued unabated. Six years later, the symptoms that warned him of an impending 30-second loss of consciousness and possible convulsion had became increasingly unpleasant. Whereas once he had just become somewhat dizzy and then hallucinated that someone was calling "Sylvère, Sylvère, Sylvère," now he was starting to feel nauseated before the onset of a seizure. The warnings still enabled him to avoid falling down and hurting himself, but, as time went by, it became increasingly clear to his doctors that his condition warranted the attention of a neurosurgeon.

Sylvère ultimately chose to undergo brain surgery, not a bad idea in his case, because one of the world's preeminent neurosurgeons of the era was then operating in the vicinity. After studying under Nobel laureate Sir Charles Sherrington, Dr. Wilder Penfield founded the Montreal Neurological Institute of McGill University. There, he had developed a surgical technique designed specifically to alleviate the symptoms of certain forms of epilepsy, and he thought Sylvère's symptoms made him a good candidate for the procedure. The strategy was to measure the electrical activity of Sylvère's brain prior to surgery, to thereby identify which area of his brain was responsible for his seizures, and then to operate to remove the identified pathological brain tissue. This might not sound like a particularly horrific plan until you hear that Penfield thought the best way to spare healthy, necessary brain tissue was to have the patient perfectly awake and conscious for the duration of the surgery.

On the day of his operation, Sylvère's head was shaved, his scalp was anesthetized, and an almost-circular incision was made along the right side of his head, from near his right eyebrow up and around to a spot behind his right ear. Then, Penfield peeled back Sylvère's scalp and just sawed right through the bone, continuing until he could carefully remove the piece of skull freed from Sylvère's head. Once Sylvère's brain was exposed beneath the breach in his skull, Penfield touched an electrode to various places in his brain, one after another, each time delivering a minute amount of electric current to that spot. The idea was to determine—by getting Sylvère's reaction to stimulation of each spot—what each of his brain areas was doing, so as to avoid damaging any that were involved in essential functions.

The range of responses Penfield recorded from his awake, conscious patients undergoing this procedure was astonishing. Some brain areas, when stimulated, caused patients to move certain parts of their bodies. Other areas, when stimulated, led patients to report that they had just been touched at particular spots. Still other areas were associated with temporary disruption of patients' abilities to understand Penfield's words or with disruption of their own speech. But among the more remarkable responses were reports of visual hallucinations, memories, or a sense of déjà vu (an illusion of familiarity). The following are some of Sylvère's responses to the stimulation of his brain:

14. "Just like someone whispering, or something, in my left ear. It sounded something like a crowd."
16. "Something brings back a memory. I could see Seven-up Bottling Company—Harrison Bakery."
18. "Someone was there in front of me right where the nurse is sitting."
19. "Someone speaking to another and he mentioned a name, but I could not understand it." When asked whether he saw the person, Sylvère replied "it was just like a dream." When asked if the person was there, he said, "Yes sir, about where the nurse with the eyeglasses is sitting over there.

Years later, Penfield wrote "I was more astonished, each time my electrode brought forth such a response. How could it be?"

When Penfield was finally confident that he could remove the brain area responsible for Sylvère's seizures without damaging other important parts of his brain, he cut out the offending tissue (an unusually large quantity in this case, some 8 centimeters long). He then replaced Sylvère's skull, and sewed his scalp back together. Within three weeks, Sylvère was discharged from the hospital in good condition, seizure free.

How are we to understand the finding that electrical stimulation of the brain can bring sensory experiences, memories, and dreamlike states to consciousness? Contemporary neuroscientists believe that our brains store traces of the events that make up our lives; it is almost as if an adult's brain is, in some respects, like an incredibly intricate—and active!—information-filled vault. And that conceptualization leads directly to the developmental question: How does the structure of a person's brain come to reflect his life's experiences? How do brains *develop?*

FROM ZERO TO SIXTY IN ONE-HUNDREDTH OF A SECOND: BRAIN CELL PROLIFERATION

Brains are made of intricate circuits of specialized cells—called neurons—that are connected to other neurons, connected to other neurons, and so on. And each circuit serves certain functions of the brain. For instance, a normal adult human can see her world because, simplifying somewhat, visual information initially received by her eyes is transmitted by neurons back to a collection of other neurons called the thalamus, which is located in the center of her brain. The neurons that transmit visual information from the eyes can do so because they are *connected* to the neurons of the thalamus. The neurons of the thalamus then transmit this information via other neurons to the rear part of the brain where they connect with neurons that are responsible for processing visual information. This part of the brain, called the occipital (pronounced ox-sip-i-tal) lobe, is a part of the cerebral cortex. The cortex is the large part of the brain that we picture when we think of a brain; it is the part that has evolved most recently and the part that contributes to uniquely human characteristics, such as speech, dance, and logic, among many other things.

The brain is the first of the body's organs to become differentiated from the rest of the undifferentiated cells of the embryo. Human brain development begins when the sheet of ectodermal cells on the outside of an embryo is induced by the mesodermal cells beneath it to curl into a tube. This fluid-filled—but otherwise hollow—tube will become the central nervous system, including both the spinal cord and the brain. Normally, as soon as the tube has been formed, a burst of cell division activity begins at one of its ends, causing a remarkable proliferation of cells that will ultimately constitute the brain. Consider, for example, the growth of the cortex. A single embryonic cell in the tube divides in half, producing two daughter cells. Each of these cells then divides in half so that there are now four cells, and so on. Nine months later, when a newborn baby is delivered into the world, she will be born with a cortex containing more than 120 billion cells, each descended from the *same single precursor cell.* * For brain cell numbers to grow so astoundingly, brain cells have to divide rapidly; at times during a fetus's last few months in utero, a quarter of a million new brain cells are being produced each minute!

There is more to creating a functional brain than simply generating billions of cells, however. It turns out that the newly created cells that make up an embryonic brain are not connected to one another at all! In fact, these cells aren't even neurons yet; instead, they are immature, undifferentiated, round cells called "neuroblasts" that lack the characteristic features of mature neurons. The early, proliferative stages of brain growth, then, merely leave behind a brain packed with unconnected neuroblasts that are ready to change with further experience.

When you start thinking about it, it is not at all obvious how the embryonic brain's new neuroblasts manage to hook up with one another in a normal, functional way. In part, this is because some cells that will be connected to each other in adult brains are physically quite distant from each other when they first begin to form their connections. This leaves Mother Nature with a problem: How can cells that need to be connected with one another find each other across great distances, distances that are in some cases thousands of times greater than the size of the cells themselves?

FINDING A NEEDLE IN A GALAXY: AXON GUIDANCE

The process of forming connections begins when a neuroblast sprouts an outgrowth—called an axon—that will ultimately extend to and connect with another developing neuron with which the cell needs to communicate. But how do these new axons find their ways to their target neurons? Work conducted by Nobel laureate Roger Sperry on the visual system of amphibians provided a potential answer to this question. Newts are able to regenerate their neurons; thus, to study this problem, Sperry cut the axons carrying visual information from newts' eyes to their brains and watched as they reformed their connections. On the basis of anatomical evidence that the severed axons had grown back to—and reestablished their connections with—their original targets, Sperry concluded that regenerated axons are *directed* to rather precise targets by a chemical affinity for some sort of distinctive, *recognizable* molecule present in their local microenvironments. In 1965, he wrote "early in development, the nerve cells, numbering in the millions, acquire and retain thereafter, individual identification tags, chemical in nature, by which they can be recognized and distinguished from one another."

The results of a 1991 study conducted by Z. Molnar and C. Blakemore supported Sperry's contention that nature solves the problem of axon guidance, at least in part, by utilizing chemical signals. Taking advantage of the amazing fact that we can now keep chunks of brain tissue alive and growing in laboratory dishes, these researchers studied the effects of juxtaposing a disembodied piece of "visual" thalamus and a similar piece of "visual" cortex, brain areas that are normally connected to one another in living people. Sure enough, when placed next to each other in a dish, the thalamic cells began to grow axons in the direction of the cortical cells. To Molnar and Blakemore, it looked like the thalamic axons were being attracted to some sort of signal arising from the target cells in the cortex.

This process is now understood to work as follows. The leading edge of a growing axon—the so-called growth cone—has tiny antennaelike elongations called "filopodia" that protrude from its surface. Filopodia are extensions of the cell membrane that literally feel their way through their microenvironments, actively sensing and drawing toward distinctive chemical cues secreted by target cells into the local environment. Because such chemical cues are most concentrated near the target and are increasingly diluted at greater distances, the developing axon can follow this concentration gradient to the target, much as you can follow a faint whiff of garlic to the fabulous restaurant that's cooking it.

Once a growth cone identifies a chemical signal coming from a particular direction, the axon's filopodia extend in that direction and latch firmly on to the surface that the axon is traversing. Then, the filopodia contract, actually pulling the axon toward the target (as this is happening, the cell produces new material that is incorporated into the cell membrane, increasing the membrane's surface area and thereby lengthening the axon). The first time I saw footage of axons seeking their targets, I was astonished; their movement appeared so purposeful! In fact, observers are often struck by the fact that they are *alive*. The brains of embryos and fetuses are literally crawling with swarming masses of neurons, each of which is alive in its own right and sensitive to its microenvironment.

FERRET FOLLIES: THE INCREDIBLE FLEXIBILITY
OF BRAIN STRUCTURE

Despite our growing understanding of axon guidance, the portrait just painted still leaves us with some big unanswered questions. This became apparent when Molnar and Blakemore repeated their brain-on-a-plate experiments with other brain regions. In their original study, the piece of visual thalamus generated axons that grew toward the piece of visual cortex. But what would happen if the piece of visual thalamus were put in a dish near a chunk of cortex that ordinarily receives, for example, auditory information? Molnar and Blakemore discovered that visual thalamic axons are not particular; hell, they'll grow toward pretty much *any* piece of cortex, regardless of what that area of cortex usually does! Clearly, the factors guiding axons are complex: cortical cells attract thalamic axons, but thalamic areas that normally process visual information need not send axons to cortical areas that normally process this same type of information. The parts of the cortex that normally process information about sounds seem perfectly able to receive information about sights instead.

But do Molnar and Blakemore's results bear at all on what happens inside a living animal? This question is not easily answered experimentally, but the available data suggest that living brains, in large part, are as flexible as were Molnar and Blakemore's in vitro brain slices. For example, in an exceptional study conducted on ferrets, M. Sur, S. L. Pallas, and A. W. Roe examined the ability of cortical cells to process information very different from what they normally process. First, operating on newborn ferrets, Sur and colleagues surgically destroyed both the visual cortex and another part of the brain that usually receives input from the eyes; together, these actions led to degeneration of the cells in the thalamus that normally receive visual input. As a result, axons coming from the eyes could not make the connections they would ordinarily have made with those (now nonexistent) thalamic cells. Then, the scientists destroyed the cells that normally send auditory information to the part of the thalamus that normally receives such input; thus, cells in this part of the thalamus did not receive their usual inputs. Under these abnormal circumstances, developing axons coming from the eyes connected with thalamic cells that usually receive auditory input (they were now the only cells in the thalamus available for connection). Because these thalamic cells then sent their output to the auditory cortex as usual, a situation had been set up in which visual information was sent from the eyes to what is normally the auditory cortex (via the auditory thalamus). What would a brain area that normally receives information about sounds do with information about sights?

Amazingly, Sur's results suggest that the auditory cortex was able to become visually responsive after this brain area was forced to experience incoming visual stimulation. Perhaps even more remarkable was the finding that features typical of the visual cortex but *never* observed in the auditory cortex—for example, the presence of a two-dimensional, topographical map of visual space—were now detected in the auditory cortex. Furthermore, additional data suggested that the ferrets were able to correctly interpret visual information as visual, even though it was being processed by the auditory cortex. While it remains to be seen if animals with rewired brains like these can use their senses to function adaptively in the world, it appears from Sur's studies that, to modify Spemann's famous declaration, we are hearing the world with brain tissue that could have been used for seeing it. These data imply that the *basic structure* of the brain, a structure that impacts all of our psychological characteristics, is profoundly influenced by developmental experiences in ways that most people would never have imagined. Such results encourage us to view with skepticism claims that the brain is innately structured with areas dedicated to specific sorts of functions. If there is a "pleasure center" or a "language center" or any other sort of "center" in the brain, it need not be because such brain structures are innately specified "in the genes."

Still, Sur's studies were conducted on ferrets! Can they really be taken to mean that *we* see with the part of the brain we see with because that is the part of the brain that has *experienced* visual input? Other data collected in the last two decades suggest that such a generalization might be reasonable. In 1983, H. J. Neville, A. Schmidt, and M. Kutas reported that people who have been deaf since birth have more sensitive peripheral vision than hearing people; importantly, stimulation of these people's peripheral vision results in significantly increased activity in their *auditory* cortex relative to that detected in the auditory cortex of similarly stimulated hearing people. This indicates that a congenitally deaf person, who has never processed sound using her auditory cortex, can instead use this part of her brain to process visual information. Neville concluded that her results "imply that early auditory experience influences the organization of the human brain for visual processing."

Similar results have been reported in studies of congenitally blind people; it now appears that blind people can use "visual" cortex to process information from other senses. Specifically, recent studies have shown that people blind from early childhood can use what is normally "visual" cortex to instead help them to (a) determine the location of a source of *sound* and (b) process *touch* information detected while reading Braille or embossed roman letters.* According to L. G. Cohen, these results "support the idea that perceptions are dynamically determined by the characteristics of the sensory inputs rather than only by the brain region that receives those inputs, at least in the case of early blindness."

Such findings are not limited to cases involving sensory handicaps. For example, in 1998, C. Pantev and colleagues reported that activity in the brains of trained musicians differs from activity in the brains of nonmusicians when both are exposed to certain musical stimuli. Specifically, these researchers found that the brains of highly skilled musicians generated a response to piano tones that was 25 percent greater than that generated by the brains of nonmusicians exposed to these same sounds. The increased brain response of the musicians might reflect their musical training; in fact, the researchers discovered a correlation between brain responses to piano tones and the age at which the musicians began studying music. As Pantev observed, "the younger the subjects started playing their instrument, the larger their cortical reorganization."*

Encouraged, in part, by these findings, K. Amunts and colleagues explored the possibility that musical training might affect brain areas involved in *controlling* the hands as much as it affects brain areas involved in *perceiving* auditory stimuli. Using magnetic resonance images, these researchers demonstrated that the brains of keyboard players have a gross structure that is significantly different than that seen in the brains of non-musicians. In particular, they found an indication of a correlation between the size of the brain area that initiates movements and the age at which musical training began; the earlier in life a child began playing her instrument, the larger the size of the brain area devoted to moving her hands. Apparently, the acquisition of musical skill is accompanied by a reorganization of brain areas devoted to both perception and movement, a reorganization that reflects "the pattern of sensory input processed by the subject during development."

Taken together, these data suggest that experience plays a major role in giving human brains the structural and functional organization that underlies low-level psychological processes such as perception and action. Brain structures underlying higher-level psychological processes (like intelligence, temperament, and personality, among others) are likely to be *at least* as influenced during development by nongenetic factors.*

THE LIFE AND DEATH OF A SYNAPSE

I have been writing as if the matter of connecting one neuron to another is relatively simple once an axon from one cell has found its target cell. As usual, the reality of the situation is significantly more complex. For one thing, by branching at its tail end, a single axon can make connections with more than one cell and can make more than one connection with a single other cell. In fact, the average neuron sends information (through connections) to about 1,000 other neurons, and it receives information (again, through connections) from another 10,000 or so. Given that our brains each contain more than 100 billion neurons, there are more connections in *each* human brain than there are stars in the Milky Way!

In order for neurons to pass signals to other cells, a connection between them must be established. The connections that are made between neurons (or between neurons and the cells of the organs the neurons control) are called *synapses*. The word "synapse" refers to the junction between cells, including the cell membrane at the end of the message-sending cell's axon, a portion of the cell membrane of the message-receiving cell, and the space between them. Because of their relative locations with respect to the space between them, the message-sending cell is known as the presynaptic cell and the message-receiving cell is known as the postsynaptic cell.

We know the most about the development of a particular type of synapse, namely the synapses between neurons carrying commands to move and the muscles that carry out those commands; other synapses probably develop in similar ways. Initially, before a neuron's axon arrives at its target muscle cell, the membrane of the muscle cell has protein molecules necessary for intercellular communication sparsely distributed across its surface. Once the axon arrives, though, the density of these proteins greatly increases in the vicinity of the newly arrived axon, in preparation for synapse formation; this is due to both increased protein synthesis and migration of proteins through the membrane toward this area. These changes result from the arrival of the presynaptic neuron's axon; without the signals that the axon releases to induce protein synthesis and migration in the postsynaptic cell, a synapse would not form. Thus, synapse formation requires a presynaptic neuron to be present in a postsynaptic cell's environment.

Once formed, though, a synapse is by no means permanent. On the contrary, normal development involves the elimination of large numbers of previously formed synapses, often as a result of experience. The development of so-called *ocular dominance columns* in the brain will serve as a prototypical example. Normally, cells carrying information from your right eye connect (via the thalamus, of course) with particular cells in your visual cortex, and cells carrying information from your left eye connect (again, via the thalamus) with *other* cells in your visual cortex. Since cortical cells that respond primarily to a particular eye remain grouped together in particular areas, a normal adult's visual cortex is characterized by alternating bands of cells, each of which receives information primarily from only one eye. These equal-width stripes are called "ocular dominance columns" (ODCs), because the single eye represented in a given column is "dominant" in that column. As we will see, the development of ODCs highlights the importance of experiential factors in the elimination of already formed synapses.

Early in postnatal life, each of the cells of a cat's visual cortex receives input from both eyes; ODCs do not exist in newborn kittens' brains. Over the next month, however, the columns appear. It turns out that ODCs are caused by the retraction of axon branches that had previously made synaptic connections with cortical cells. The pruning of synapses that results is a consequence of an experience-based competitive process that occurs among axons.* We know it works this way because of experiments on kittens who were deprived from birth of stimulation in just one eye. By the time they were three months old, the brains (and behavior) of these cats clearly showed the effects of their experiences. In particular, besides being functionally blind in their deprived eye, their brains had abnormal ODCs: the columns receiving information from the normal eye were wider than usual, and the columns receiving information from the deprived eye were extremely narrow (that is, few cells received input from the deprived eye). What this means is that the equal-width bands seen in normal, two-eyed cats' brains must result from the *experiences* these animals have as they develop, namely experience with equivalent visual input received in each of the two eyes.*

One of the more interesting aspects of this finding is that in the absence of experimental manipulations, ODCs in cats would *seem* to develop independently of experience, because they characterize *every normal* cat, whether reared in the wilds of New Guinea or in a posh London apartment. Because normal cats *always* experience simultaneous binocular (two-eyed) stimulation early in development, ODCs characterize normal feline brains just as surely as whiskers characterize normal feline faces. But even though the characteristic structure of this part of a cat's brain is seen in all normal cats, "information" about this structure is *not* present in the cat's genes. Instead, this structure results from species-typical interactions between biological components of normal cats and experiences that are so common in feline development that they are practically universal, and so effectively *expectable*. Recognizing this phenomenon, William Greenough and colleagues pointed out in a landmark 1987 paper that some developmental processes—they called them *experience-expectant processes*—capitalize on environmental regularities by utilizing

> environmental information that is ubiquitous and has been so throughout much of the evolutionary history of the species. Since the normal environment reliably provides all species members with certain experiences . . . many mammalian species . . . take advantage of such experiences to shape developing sensory and motor systems.

Thus, because certain *experiences* have consistently characterized the development of certain species throughout their evolutionary histories, these animals have come to require these experiences for their normal development. In this way, organisms can reliably develop traits that *appear* innate until we look more deeply into their developmental histories.

In interpreting findings on the effects of experience on the development of normal, species-typical structures in sensory cortex, Mark H. Johnson wrote, "it is almost as if the sensory surface *imposes* itself on to . . . the thalamus, and finally on to the cortex itself. . . . There is little evidence that [some normal, species-typical brain structures] are prespecified." This understanding of the development of some of the basic structural characteristics of our brains is now well accepted. In the words of the Nobel

Prize–winning neurobiologist Gerald Edelman, "the principles governing these [developments] are epigenetic. . . . The connections among the cells are . . . not precisely prespecified in the genes of the animal." As a result, "even in genetically identical twins, the exact same pattern of nerve cells is not found at the same place and time." Thus, structural characteristics typical of mammalian brains are not determined by genetic factors alone (although these certainly have their imperative influences). Instead, at the most rudimentary level, the development of our brains—the source of all of our behavioral, cognitive, and emotional characteristics—depends unequivocally and integrally on our experiences.

"PROGRAMMED" CELL DEATH

Synapses are not the only features of brains that are pruned as a result of experience—entire neurons are created only to die shortly after making connections with their targets. As paradoxical as it sounds, the most dramatic changes in our brains as we grow from birth to adulthood actually involve the adaptive *death* of nearly 20 billion cells in the cerebral cortex.

The first evidence to shed light on the mechanism underlying this phenomenon was gathered in 1909, after M. L. Shorey experimentally amputated limb buds from chick embryos and amphibian larvae (limb buds are the protrusions in embryos and larvae that normally develop into limbs). Specifically, he found that limb bud removal resulted in animals with spinal cords that contained fewer neurons where there normally would have been neurons connected to the now-missing limbs. On the flip side, grafting an extra limb bud onto a chick embryo leads to the survival of *more* neurons. Thus, there is a strong correlation between the amount of target tissue available for neurons to connect with and the number of neurons that are present in the mature animal. These results could mean one of two things: either the amount of target tissue somehow affects the number of neurons produced by cell division early in the embryonic stage, or the amount of target tissue somehow affects the ability of already-produced neurons to survive. V. Hamburger's demonstration that target tissue amounts do not affect the initial proliferation of neurons—the same number are produced under both normal and abnormal conditions—means that, somehow, the removal of target tissue must contribute to the *death* of cells that have already developed. The consensus view of neuroscientists today is that neurons are somehow *supported*—or, in effect, nurtured—by the target tissues they encounter in their local environments.

Thus, nature has generated a remarkable solution to the problem of how to give the brain control over muscles and other organs: produce more neurons than necessary and then prune them back until there are just enough for the amount of muscle (or other organ) that needs to be controlled. So, even though it sounds wasteful to kill as many as two thirds of the neurons originally produced (as happens in some brain areas), such cell death is actually *a good thing*; it leaves us with just the right number of neurons—and just the right specific neurons—to control our muscles and other organs and thereby function adaptively in our environment. (Interestingly, this arrangement also means that if an evolutionary event some day furnishes one of us with four arms, the basic mechanisms for giving the brain control of these new limbs are already in place.) But how does a developing body "know" which neurons it needs to control its muscles and organs, and which are expendable?

It doesn't. Nature has solved this problem, ingeniously as usual, with chemicals called *trophic factors*. Our understanding of the role of these chemicals in neuron death and survival grew from the Nobel Prize–winning research of Rita Levi-Montalcini, who at the time was following up on some unexpectedly oddball findings reported by other researchers in Hamburger's laboratory. These scientists had found that when they removed a tumor from a mouse and transplanted it into a chicken embryo (how's that for truly weird science?), neurons that were developing in the embryo near the tumor grew much larger than normal. Levi-Montalcini and colleagues managed to demonstrate that the tumor was releasing a substance that contributed to the survival and growth of the neurons; they called it nerve growth factor (or NGF), the first trophic factor to be discovered. Eventually, in an effort to purify and chemically identify the substance, they used snake venom to break down certain chemicals in the mix they were testing, the idea being to rule out those chemicals as the source of the growth phenomenon. To their great surprise, they discovered that the venom itself was an even richer source of NGF than the tumors they were using! From there, it was a short leap to their discovery that the salivary glands of mice were an even richer source of NGF. And once such a copious supply could be obtained, chemical analysis of NGF became possible.

Ultimately, it was discovered that NGF is a diffusible protein that is secreted by all relevant target cells and that *must* be taken up by the approaching axons of certain neurons if these neurons are to survive. We know this because injections of NGF into embryos produce animals with more neurons than usual, whereas daily injections of NGF antiserum into newborn rodents produce animals with an obvious loss of neurons. The available experimental data suggest that distinctively shaped protein receptors for NGF are embedded in the membranes of presynaptic neurons. When NGF is released by a target (postsynaptic) cell into a functional synapse, it binds with these receptors and is subsequently absorbed into the presynaptic neuron and transported back to that neuron's nucleus. In this way, the NGF effectively "informs" the presynaptic cell that a functional synapse has been formed with the target cell.* Any neuron that does *not* take in enough NGF (or some other trophic factor) from its environment will self-destruct shortly thereafter.*

The upshot of this arrangement is that neurons wind up effectively *competing* with one another for limited supplies of trophic factor, and, as a result, the only neurons to survive are the ones that successfully form working synapses with their appropriate target cells. Thus, if some environmental agent interferes with the development of a fetus's limbs, the absence of normal amounts of muscle in the microenvironments of the neurons that would ordinarily connect with those muscles ensures that the mature nervous system will reflect the absence of the limbs. Such situations clearly show the effects of environmental factors on the number of neurons that ultimately come to characterize the human nervous system.

The use of trophic factors to support the survival of only those cells that form functional synapses gives the developing brain a remarkable ability to structure itself in accordance with its environment. Although no one yet understood *how* brains are able to self-organize in an environmentally sensitive way, this ability became apparent three decades ago when Blakemore and G. F. Cooper reared newborn kittens in visually impoverished environments. It turned out that by the time they were four months old, kittens reared in an environment containing *only* horizontal black and white stripes developed into cats that did not respond to vertical lines. In contrast, kittens reared in an environment containing only vertical stripes became unresponsive to horizontal lines. As a result of being reared in such environments, these cats wound up both behaviorally and neurologically abnormal; kittens who had never seen vertical lines were liable to walk into the (vertical) legs of chairs, and brain cells that normally respond to vertical lines did not respond to such lines in these cats. Thirty years ago, standing at the scientific frontier of their time, Blakemore and Cooper took their results to mean that "the visual cortex may adjust itself during maturation to the nature of its visual experience." Today, neuroscientists understand that the cortex does adjust itself to the nature of its experience, and in more than just the visual mode; it has become clear that the mature structure of the cortex cannot possibly be predetermined by genes. Taken together, these findings reveal just how dependent the normal structure and function of our brains is on nongenetic—in this case, macroenvironmental—factors.

As we develop, microenvironmental and macroenvironmental factors importantly influence how our brains are built. Cells differentiate under the influence of neighboring cells. Target cells secrete chemicals into their microenvironments, thereby guiding axons to their destinations. Binocular experience with visual information from the macroenvironment gives rise to the characteristic "striped" structure of the visual cortex. A neuron lives or dies depending on its ability to absorb trophic factors that its target releases into its microenvironment. And the stimuli we encounter in the external world—and the ways in which we behave—importantly influence the size and functioning of the sensory and motor areas of the brain. On both macroscopic and microscopic levels, then, the environments in which we develop powerfully influence the structures and functions of our brains.*

Part III

DEVELOPMENTAL SYSTEMS

To counteract ignorance we should concentrate
on . . . interdependence.

—*Tenzin Gyatso, the fourteenth Dalai Lama*

7

A TURTLE IN THE SHADE
The Development of Sexual Characteristics

Beaming, my beloved mechanic, Nick, recently announced the birth of his third daughter. (I began driving an Alfa Romeo years ago in graduate school, and when you're committed to a temperamental Italian roadster, you quickly learn to cherish a competent Italian mechanic.) Before the birth, Nick had thought it might be nice to finally have a son, but apparently that was not meant to be. When I joked that Nick should have asked his wife, Beverly, to try some of the various ploys that have been rumored to influence a baby's sex, Nick told me that Beverly had been rather insistent with him: the sex of a baby is determined by its father. Ever the instigator, I told Nick to tell Beverly that sex determination is actually quite a bit more complicated than that. But when my muffler fell off a few weeks later and I was back at Nick's garage, he told me that Beverly would have none of that: the sex of a baby is determined by the father, and Nick alone was to credit for the sex of his new baby. Fortunately, Nick seems happy to claim responsibility at this point; the arrival of his new daughter has left him with a smile indelibly plastered across his face.

Much of the public today believes that sex is determined almost single-handedly by the presence or absence of a Y chromosome in an embryo; the presence of this chromosome is understood to produce a male while its absence is understood to produce a female. And since only a man's sperm can contribute a Y chromosome to a zygote, then, according to the genes-determine-sex idea, Beverly was right: fathers determine the baby's sex. But while it is true that in humans, a particular "gene" on the Y chromosome can initiate a complex series of events that usually leads to the masculin-ization of the embryo, the notion that sex is strictly genetic remains a mis-leading simplification.

Consider the observation that in several nonmammalian species, sex-ual traits can be radically affected by macroenvironmental events; this fact highlights the ability of nongenetic factors to influence the ultimate form of traits essential to an animal's identity. For example, there are no *genetic* differences at all between turtles of different sexes, but mature males and females are easily distinguished both by the appearance of their bodies—males are much smaller than females—and by their behavior. As is the case with some species of lizards and with all alligators and crocodiles, the *temperature* in a turtle egg's local environment is what "determines" the developing turtle's sex. For many species of turtles, newborns that hatch from eggs laid in the sun (and that subsequently develop in environments warmer than 32 degrees Celsius) are female, whereas newborns that hatch from eggs laid in shadier areas (and that develop in environments cooler than 28 degrees Celsius) are male.*

Other environmental stimuli that can influence sexual differentiation include physical position or social standing. A worm called *Bonellia* becomes male if it develops in a particular place, namely the proboscis (a worm's "mouth") of another *Bonellia*; otherwise, it develops into a female (males spend their entire lives inside the female, fertilizing her eggs). Many species of fish actually retain the ability to change their sex throughout the course of their lives. In particular, numerous species of coral reef fish can, as a function of their social environment, switch from female to male, a conversion that requires significant bodily as well as behavioral changes. In many such species, a large male will have a harem of females that he dominates, defends, and mates with. If this male dies or is removed from the population, the next largest individual—a female—undergoes a spontaneous sex change within a matter of hours, taking over the role of the large male. Within days, this previously female fish now produces sperm, behaves like a male, and develops the distinctive coloration of the male, effectively changing its appearance entirely. While mammals cannot (as far as we know!) spontaneously change sex as a result of social stimulation, this example illustrates how bodily characteristics such as coloration, behavioral characteristics such as sexual activities, and sexual characteristics like the ability to produce sperm can be *open* to environmental stimulation, even in some multicellular vertebrate animals.

Given data on the development of sexual traits in coral reef fish, turtles, and crocodilians, one might be tempted to conclude that in these species, sex is *determined* by the environment (using a hazardous linguistic shorthand, some developmental biologists have done so). After all, there are no genetic differences between male and female crocodiles, and whether a crocodile develops into a male or female depends on an environmental factor, namely incubation temperatures. Thus, to hearken back to the terminology of behavior genetics, *all* of the normal variation in sex characteristics in these animals is "accounted for" by variation in environmental factors. But does this mean that a crocodile's sexual traits are *caused* by the environment in which the crocodile develops?

Of course not. Ultimately, it makes no sense to argue that a crocodile's sex is caused by its environment, because without the crocodile's genes, there would be no crocodile at all, let alone one of a particular sex! Clearly, the environment does not contain all of the "information" needed to construct crocodile sexual organs, so there is something misleading about the claim that the environment alone is responsible for a crocodile's sex. Turtle, worm, or fish zygotes without the genetic machinery to *respond* to environmental cues such as temperature, location, or social standing, cannot possibly develop into turtles, worms, or fish with particular sexual traits. Likewise, human embryos that develop in environments lacking *non-genetic* factors that respond in particular ways to the activity of the genes cannot possibly develop into boys or girls. So, while there is a sense in which a crocodile's sex is determined by its environment—the same sense in which a human baby's sex is determined by its father—there is a more important sense in which it is not. The claim that the environment determines the sex of a crocodile ultimately confuses our thinking about the causes of crocodiles' sexual traits; the claim that mammalian sex is determined by genes alone is equally obfuscating. As is the case for all traits, sexual characteristics in mammals and nonmammals alike are determined by complex epigenetic interactions of genetic *and* nongenetic factors.

Sexual Traits and the Hormonal Microenvironment

In mammals, *genetic sex*—by definition—is determined by inherited chromosomes. There are two forms of the human sex chromosomes, called "X" and "Y," respectively. Normally, each cell in a male's body contains one X and one Y chromosome, whereas each cell in a female's body contains two X chromosomes. In contrast to genetic sex, the development of most male and female *traits* is influenced by hormones, substances secreted from glands into the blood and then carried by the bloodstream to the specific target cells they affect. According to L. Wolpert:

> The only cells that are affected by the sex chromosomes are the germ cells [that is, sperm and eggs] and those in the tissue that will give rise to an ovary or a testis. All the main sexual characteristics that make males and females different arise from the effect of the hormonal secretions of the testis.

An important part of understanding sex determination in mammals, then, is understanding the origins and effects of these hormones.

The geneticist Alfred Jost wanted to understand the workings of sex hormones, and he wasn't going to let the fact that fate gave him extraordinarily poor working circumstances interfere with his experimental investigations. Working with rabbits in Nazi-occupied France near the end of World War II—a state of affairs that sometimes required him to sleep with his furry subjects at night, since this was the only way to keep them from freezing to death—Jost attacked his problem. His great breakthrough in understanding came from studies in which he surgically removed from *fetal* rabbits the tissue that would ordinarily have developed into gonads (testicles, or testes, in XY individuals, and ovaries in XX individuals). In every case, regardless of the genetic sex of the rabbit, a female rabbit was ultimately born, complete with uterus, cervix, and vagina. The finding that females develop in the absence of any gonads led Jost to conclude that the female form is the "default" form of mammals, the form that develops in the *absence* of a masculinizing factor, regardless of whether or not any sort of feminizing factor might be present.

Given that the presence of a testis was the only difference between his normal genetic males (who looked male) and his "castrated" genetic males (who looked female), Jost reasoned that a factor produced in the male gonads must be necessary for the development of normal masculine traits. Scientists now understand that in mammals, this factor is the steroid hormone *testosterone*, one of the class of masculinizing hormones called "androgens."* The masculinizing effect of testosterone can be demonstrated experimentally by injecting it into pregnant guinea pigs; as a result of this treatment, offspring that are genetically female (XX) develop external genitalia that are identical to those of normal males. Thus, a Y chromosome is not required for the development of male external genitalia.

But if a Y chromosome is not necessary for the development of such essentially male traits, why is the presence of these traits in human beings so highly correlated with the presence of a Y chromosome? The answer is that a minute portion of the human Y chromosome—called the *SRY gene*—can contribute to the production of *testis-determining factor*, a protein that normally contributes to the development of testes in XY embryos that are more than two months old (younger "male" and "female" embryos are indistinguishable from one another). The absence of this protein permits the development of ovaries. And it is the testes or ovaries that then bathe the fetus in the hormones necessary for the development of other sexual traits; testis-produced testosterone normally contributes to the development of masculine traits and ovary-produced estrogen (the class of feminizing hormones) normally contributes to the development of feminine traits. Thus, an SRY gene cannot single-handedly produce a hairy chest, a deep voice, baldness, or a beard; as a portion of a chromosome, it can do only what "genes" do, namely produce a protein. The question, then, becomes how can the production of this single protein have such far-reaching consequences?

It turns out that the unique shapes of some proteins give them the ability to turn certain genes "on" and "off." That is, some proteins (or groups of proteins) are shaped in such a way that—like an appropriately shaped key in the right lock—they can *activate* some genes, causing those genes to produce the proteins they code for. Similarly, some proteins (or groups of proteins) can *inhibit* the activity of some genes, preventing them from "expressing" their protein products. This arrangement has incredibly important implications: since genes produce proteins, and since some proteins can activate or inactivate other genes, then *some genes can effectively control other genes via their own protein production.* *

Recent research suggests that testis-determining factor—the protein the SRY gene codes for—*probably* controls other genes that produce proteins needed for the collective construction of a testis. But for the moment, this remains a speculation. As Gilbert notes:

> We still do not know what the testis- or ovary-determining genes are doing. . . . The problem of primary sex determination remains (as it has since prehistory) one of the great unsolved problems of biology . . . [partly because the SRY gene] is necessary but not sufficient for the development of the mammalian testis.

Thus, to bring about the development of a testis, SRY has to work cooperatively with other genes located on non-sex chromosomes. The fact that a Y chromosome alone cannot cause the development of male traits is underscored by the observation that in rare cases, people *with* an SRY gene can nonetheless develop female traits, while people lacking an SRY gene can nonetheless develop male traits.

Actually, the development of male traits is not even an inevitable result of fetal exposure to testosterone. In a disorder known as *androgen insensitivity syndrome*, genetic males (XY) can be born with (and develop into adults with) an external appearance that is indistinguishable from normal females. While these so-called pseudohermaphrodites look exactly like normal women, they have a Y chromosome in every cell of their body, they have (internal) testes, and they are sterile (lacking, as they do, uteruses and oviducts). Here's what's going on: because they have a Y chromosome, their testes initially develop normally, leading to the subsequent production of testosterone. But because of their disorder, their cells are deficient in the production of *receptors* for androgens, so they cannot respond to the hormone bath in which they develop. Therefore, to develop into a normal male, it is not enough to be exposed to testosterone (let alone to have a Y chromosome!); particular cells in one's brain and body must be able to both *recognize and respond* to the testosterone as well.*

Thus, while the SRY gene is *normally* associated with the presence of testis-determining factor—which, in turn, is normally associated with the presence of testes, testosterone, and masculine traits—SRY clearly *does not directly cause* the development of male traits. Instead, these traits result from a complex chain of events that flow, one after another, in a history-dependent cascade that is open to microenvironmental influence at many places along the developmental pathway. And keep in mind that I have been discussing in this chapter only the most purely "biological" aspects of our sexual identities. Obviously, social factors, too, contribute to the masculinity and femininity of the people around us. But if our "biological" sexual traits are not determined by our sex chromosomes alone, it is inconceivable that the *psychological* traits associated with men and women can be considered "genetically determined." The take-home message is the same for all of our traits: traits are not caused by genes alone but instead are caused by a complex variety of interacting genetic and nongenetic factors, all of which affect development.

THE GENDERED BRAIN

The effects of hormones on developing brains have been studied in great detail. Scientists now know that exposure to steroid hormones has an *organizing* effect on the fetal brain, such that prenatal exposure to testosterone normally results—after development to maturity—in a male brain that continuously and steadily secretes certain gonad-stimulating hormones. In contrast, lack of prenatal exposure to testosterone normally results in a mature female brain that secretes these same hormones in cyclical spurts, thus giving rise to women's monthly periods. But how can *prenatal* exposure to these relatively simple molecules have such incredibly far-reaching consequences so many years after the developing brain is first exposed to them?

The answer to this question is similar to that offered above for testis-determining factor. Steroid hormones (including testosterone, the estrogens, and cortisone) have the effects they do because they, too, can turn genes "on" or "off." In fact, steroid hormones can diffuse right across cell *and* nuclear membranes and float directly into a cell's nucleus. There, they can bind with uniquely shaped protein receptors that "recognize" the hormones by their distinctive shapes, using a "lock and key" type of strategy (these are the receptors that are lacking in individuals with androgen insensitivity syndrome). The steroid-receptor *complex* then becomes able to bind with DNA; thus, just like proteins such as testis-determining factor, the steroid-receptor complex can turn certain genes "on" or "off," or regulate the *rate* at which they are decoded. This ability of hormones to turn genes on and off has staggering ramifications that I will discuss further in later chapters. For the moment, it is important to note only that this arrangement provides a mechanism whereby certain macroenvironmental experiences can effectively control the activity of certain genes.

Given the ability of hormones to turn genes "on" and "off," it is perhaps not surprising that they have profound effects on developing brains. A detailed examination of the hormonal masculinization or feminization of the brain is beyond the scope of this book, but before leaving the topic, I must mention one particularly important effect that hormones can have on neurons. Gonadal steroid hormones can affect the onset and growth rate of both axons and dendrites (dendrites are projections that sprout, axonlike, from the nonaxon end of neurons and that receive incoming information from other neurons' axons). Cells containing appropriate receptors respond to testosterone or estrogen both by producing new outgrowths and by causing extensive new branching of already existing axons and dendrites. Thus, like NGF, these hormones have powerful effects on the growth, development, and structure of the brain.

SEXUALITY AND THE PRENATAL ENVIRONMENT

So far, I have been describing only *micro*environmental influences on the development of our sexual characteristics; each of the factors I have mentioned operates *within* the body of the developing person. As is often the case, though, there are also factors outside a developing person's body that can affect trait development. It turns out that one such effect was hidden in Jost's finding that female traits develop in "castrated" rabbit fetuses. How and why do *female* traits develop in animals unexposed to androgens, *even if these animals are also unexposed to feminizing factors like estrogen?*

The answer has to do with the fact that mammalian fetuses are always carried in the bodies of their mothers, not their fathers. Because of this arrangement, *all* mammals develop in an environment containing feminizing hormones. We can see the effects of developing within a woman's body by considering a human disorder called Turner's syndrome. This condition develops when one parent contributes an X chromosome to a zygote and the other parent contributes no sex chromosome at all. In such cases, normal ovaries do not develop, so the fetus itself cannot be a source of estrogen. Nonetheless, like Jost's "castrated" rabbits, Turner's infants are born with female genital tracts; they are so similar to normal girls that they sometimes go undiagnosed until adolescence.* The available evidence suggests that female genitalia develop in these individuals as a result of exposure to the maternal- and placenta-derived estrogen in the fetus's macroenvironment. Thus, in this case, the fetus *is* feminized by maternal and placental estrogen *located outside of its body.*

Similar effects of hormones originating outside the developing body have been demonstrated in other mammalian species. I have a colleague with a female dog who urinates by lifting her leg in a stereotypically male way. Similarly, my mom's female dog has developed the strange habit of trying to "mount" my dad's elbow, much as male dogs mount female dogs. What could explain such unusual—and, to everyone besides my dad, hilarious—behaviors?

R. L. Meisel and I. L. Ward have provided a possible answer to this question by discovering why some otherwise normal female rat fetuses develop into rats that behave in some stereotypically male ways (for example, by mounting other females or by being unusually aggressive). It turns out that fetal rat littermates share maternal blood flow, and that female rat fetuses "downstream" from fetal brothers are the ones most likely to develop stereotypically male traits. Meisel and Ward concluded that a masculinizing hormone—presumably testosterone—produced by an upstream brother can be carried via the maternal bloodstream to female siblings, increasing the likelihood that these siblings will develop some stereotypically masculine traits. The presence of brothers in utero seems to affect males as well; in an article entitled "Stud Males and Dud Males," M. M. Clark, L. Tucker, and B. G. Galef reported that male gerbils who developed as fetuses between two brothers sired an average of 28 percent more offspring than males who developed between two sisters. These findings might reflect the fact that female gerbils in "heat" spent significantly more time near the former, "studlier" males.

This raises the question of whether or not similar effects might occur in situations where human fetuses share a uterus. In fact, such effects would be possible for two reasons. First, it is known that testosterone injected into a pregnant monkey's bloodstream can pass through the placenta and masculinize an XX fetus. Second, it is known that blood testosterone levels in women pregnant with male fetuses are higher than those in women pregnant with female fetuses, indicating that testosterone produced by a fetus can cross the placenta and enter the mother's bloodstream. Thus, steroid hormones are passed bidirectionally between mothers and their fetuses. Given these findings, it is likely that such hormones can also be passed between fetal twins.*

The hypothesis that human hormones produced by one twin can influence the development of the other twin has not yet been proven, but the first study to look for such an effect found support for this idea. Specifically, S. M. Resnick, I. I. Gottesman, and M. McGue reported that girls with twin brothers scored higher than girls with twin sisters on a reliable measure of willingness to engage uninhibitedly in social and sexual activities. Given that average males in many cultures score higher than average females on this scale, this finding reflects the presence of a masculinizing influence on girls with twin brothers, one that does not similarly influence the development of girls with twin sisters. While this study did not rule out the possibility that the measured differences reflected the *postnatal* presence or absence of a brother, the results are consistent with the idea that sharing a uterus with a brother can affect the development of a girls' traits, presumably via exposure to brother-produced testosterone.

Usually, though, a human fetus develops in its own private uterus. In such circumstances, is exposure to hormones so well controlled that its effects are predictable and therefore, not worth discussing? Not at all.* In fact, the experiences of pregnant rats can affect both their hormonal states *and* some of the sexual traits of their offspring; such findings suggest that this might be the case for humans as well. In particular, studies of pregnant rats subjected to stress in the third week of their pregnancy found their male offspring to be more likely to "show a persistence of female behavioral potentials and an inability to exhibit normal male copulatory patterns in adulthood." The authors of these studies concluded that "the processes involved in masculinization . . . appear to have been compromised in the male fetuses of stressed mothers." Later studies demonstrated that stressing pregnant rats reduces testosterone concentrations in their male fetuses, and that this reduction can occur in specific periods critical for sexual differentiation of their brains. I am aware of only one study of this phenomenon in human beings, and it revealed that "average stress severity [scores were] . . . nearly twice as high [through months four, five, and six of pregnancy] for mothers of male homosexuals as for mothers of male heterosexuals." These results must be considered with caution, however; the authors of the report concede that their study was exploratory and fraught with methodological obstacles.* Nonetheless, these results are consistent with the hypothesis that fetal development is subject to the influences of environmental stimulation, even if these influences are not in the fetus's immediate environment but are present in the environment of the fetus's mother.

The take-home point of these last few examples warrants highlighting even though it might now seem obvious: *embryos and fetuses develop in a macroenvironment that can influence their traits.* Very often, when considering the influences of nature and nurture on a trait, we imagine that nurture does not begin until birth; to the extent that the environment of a fetus— that is, a uterus—is biological, we are tempted to count it as an aspect of nature.* Nonetheless, this environment clearly exists outside of the fetus's body, and it clearly contains influential factors that can be independent of the fetus's genes. Consequently, effects of these factors *cannot* be considered "genetically determined" by any stretch of the imagination. In addition, fetuses cannot be thought of as being insulated from the macroenvironments of their mothers; after all, the chemical state of the mother—who *is* the fetus's macroenvironment—is affected by all sorts of factors, from diet and stress to drug consumption and exposure to pollution. In thinking about the relative contributions of genetic and nongenetic factors to trait development, one must never lose sight of the fact that *development always occurs in some environment*. And, of most importance, we now know that mechanisms exist whereby that environment can affect trait development, either directly or by influencing genetic activity itself.

8

THE DUTCH HUNGERWINTER
AND THE CAT IN THE HAT
How Prenatal Environments Affect Our Traits

In October 1944, in a repugnant attempt to undermine resistance activities in northern Holland, Nazi forces began a blockade of ships carrying food and other supplies to the large cities in the western part of the Netherlands, including Amsterdam, Rotterdam, and The Hague. By February 1945, the situation had deteriorated severely, so much so that the average food consumption per person had dropped to about half of normal. This severe famine, subsequently known as the Dutch Hungerwinter, led to the deaths of thousands of people by starvation. Those who survived did so by consuming potatoes, bread, tulip bulbs, and those foodstuffs they were able to obtain by foraging in rural areas for *anything* of nutritional value, including domesticated animals such as cats. And then, as abruptly as it began, the famine ended when Allied troops arrived in May 1945.

As tragic as it may be, this tale effectively describes an unusually well-controlled—if dreadfully destructive—field "experiment": well-fed people from a variety of social classes were forcibly put on a severely restricted diet for a sharply delimited period of time, after which they quickly reverted to their prefamine diet. And while the incidence of stillbirths and infant mortality was sharply higher in Holland that winter, many babies who were born did survive. This situation has allowed researchers to study the effects of developing in a mother who experienced severe malnutrition during a very specific portion of her pregnancy.

Studies of mothers who were malnourished during the first three months, or *trimester*, of their pregnancies have revealed an increased incidence in their offspring of gross central nervous system abnormalities, including spina bifida and cerebral palsy. These studies, in concert with others, demonstrated that women in early pregnancy must consume adequate amounts of the B vitamin folic acid to ensure normal nervous system development in their children. Normal prenatal care in industrialized nations often now includes education—with dietary supplements, if necessary—about the importance of folic acid to early fetal development.

Given the increased incidence of nervous system abnormalities in these populations, it is perhaps not surprising that researchers have documented a similarly increased incidence of schizophrenia in the offspring of malnourished mothers. Specifically, compared to women whose mothers were not malnourished, women whose mothers experienced famine during the first trimester of their pregnancies were more than twice as likely to ultimately develop schizophrenia. This supports the widely held notion that prenatal factors can have significant impacts on later mental functioning. But since it is now virtually common knowledge that certain prenatal experiences—exposure to alcohol, for example—can have deleterious impacts on development, I will not focus on this sort of phenomenon; it seems fairly obvious that prenatal factors can cause gross birth defects, be they physical or psychological. What might be more surprising is the recent suggestion that prenatal factors can also affect the development of other, less obviously abnormal traits, such as obesity. Data supporting this suggestion indicate that prenatal experiences can impact brain development in such a way that the effects of these experiences are detectable many years later in otherwise normal adults.

In 1976, G. P. Ravelli, Z. A. Stein, and M. W. Susser reported the results of a study of 300,000 men born to women who were pregnant during the Dutch Hungerwinter. These researchers discovered that adult men born to women malnourished during the first two trimesters of their pregnancies were more likely to be obese than were men in the general population. Given that the offspring of mothers who were underfed for the *entire* duration of their pregnancies are known to be permanently *under*weight, this finding was remarkable indeed.

To explore this phenomenon further, Alan Jones, one of my colleagues at Pitzer College, malnourished pregnant rats for the first two-thirds of their pregnancies; if their offspring were obese as adults, Jones could then use the rats as an "animal model" of the Dutch Hungerwinter effect in humans. In their initial studies, Jones and M. I. Friedman fed pregnant rats 50 percent of their normal rations for the first two-thirds of their pregnancies and then allowed them to eat freely for the final trimester. These rats ultimately gave birth to pups that had body weights that were the *same* as those of pups born to normally fed mothers. However, weeks later, after weaning onto a high-fat diet, the male—but not female—offspring of the malnourished mothers ate more and gained more weight than did the offspring of the normally fed mothers (even though the offspring of the normally fed mothers were weaned onto the same high-fat diet). Jones and

Friedman reported that the fat cells of the obese males were larger, and that their fat pads weighed from two to three times more than the fat pads of the normal males. Apparently, they concluded, "male offspring of underfed mothers deposit excessive amounts of fat."

In the ensuing 15 years, Jones has relentlessly pursued a series of subtle scientific clues in an attempt to discover the cause of this effect, and he has recently generated a noteworthy hypothesis to explain it. In initiating his search, he reasoned as follows. First, male offspring of rats who are undernourished in early pregnancy and then fed normally during their third trimester become obese. Second, the offspring of rats who are undernourished *throughout* pregnancy are permanently underweight. Therefore, the mother's food consumption during her third trimester must somehow influence her mature offspring's body weight. Thus began the search for a biological substance, levels of which vary as a consequence of food consumption.

Insulin is a hormone that is secreted by the pancreas in response to the high blood sugar levels that normally occur after meals; it promotes the transfer of blood sugar from the blood to other cells of the body, where the sugar provides needed energy. (Diabetes is a disorder in which the pancreas does not produce insulin; a diabetic must inject a quantity of insulin into her bloodstream after meals to support the transfer of blood sugar to the cells.) Thus, blood levels of insulin are higher after food is consumed than at other times, leading Jones to hypothesize that final-trimester insulin levels might be responsible for the Dutch Hungerwinter effect. The hypothesis appears quite reasonable when you consider that insulin plays an important role in the regulation of every normal adult's body weight.

In an experimental test of the insulin hypothesis, Jones and M. Dayries injected *normally fed* pregnant rats with insulin every day during the third trimester of their pregnancies. At birth, the pups of these mothers were indistinguishable from the pups of uninjected mothers. However, once they were mature, male offspring (but not female offspring!) born to insulin-injected mothers were significantly heavier than were their insulin-unexposed counterparts. Furthermore, male rats born to insulin-injected mothers gained more weight per gram of consumed food than did their counterparts. That is, males born to insulin-injected mothers were significantly more *efficient* in their use of food. (In our overweight and weight-conscious society, we might not be inclined to use a positive word like "efficient" to refer to animals who easily convert food to fat, but this is the word biologists use.) How can insulin produce effects like this?

Jones has proposed that prenatal exposure to *metabolic* hormones such as insulin might organize the fetal brain much as prenatal exposure to the *steroid* hormone testosterone does. If this is right, then a mother's blood level of insulin might affect the *brain* of her offspring in ways that are not apparent at birth but that are seemingly permanent in adulthood. Is this possible?

In fact, insulin has several important effects on the nervous system, meaning that its presence during prenatal development could have the profound long-term consequences Jones has hypothesized. In addition to speeding up the formation of synapses in neural tissues, insulin can also stimulate the growth of both dendrites and axons, and can support the survival of neurons that would spontaneously die in its absence. Thus, insulin has many of the same sorts of effects as steroid hormones (which also stimulate the growth of axons and dendrites) and as nerve growth factor (which also stimulates axon growth and supports neuron survival). Given these facts, Jones and Dayries concluded that fetal insulin levels might be able to influence the development of brain areas that are "relevant to body weight and food intake regulation."

So far, the available evidence seems to be in line with Jones's theory. First, Jones and D. H. Olster recently discovered that malnutrition and subsequent refeeding does, in fact, lead to unusually high insulin levels during the third trimester of pregnancy. And while it is not yet known for certain if insulin crosses the placenta during this period, available data suggest that it probably can. Second, the number of insulin receptors in the fetal brain peaks in the third trimester, and these receptors are particularly concentrated in the areas of the brain that have roles in feeding behavior (the hypothalamus) and in the processing of information about smell (the olfactory bulbs). Thus, it is beginning to look like a pregnant woman's nutritional experiences at specific times during her pregnancy can affect her hormone levels in ways that profoundly influence the structure and functioning of her fetus's brain. Moreover, the available evidence suggests that this influence can continue to affect some of the developing person's traits for years to come, sometimes in ways that would have previously been unthinkable.

No one yet knows if such effects on an adult's body weight might somehow reflect an evolutionary adaptation of sorts, but the possibility cannot be ignored. Since addressing this question requires rank speculation, it is not easy to find scientists willing to step into the fray. Nonetheless, when I recently questioned Jones in an informal environment about the possible evolutionary significance of his Dutch Hungerwinter effect, he was willing to entertain this important possibility. Specifically, he spoke of the evolutionary advantages of a mechanism that would allow a fetus to effectively *sense* the nutritional nature of the environment into which it would soon be born and to use this "information" to alter its metabolism to most efficiently function in that environment.* A fetus soon to be born into an environment in which the availability of food swings rapidly from one extreme to another would be at an advantage if it could somehow "detect" this characteristic of its future environment and then alter its brain (and therefore, its future metabolism) accordingly. In situations where food availability swings rapidly from feast to famine, it is useful to have a body that efficiently stores nutritional reserves as fat during times of feast, to be accessed and utilized later, during times of famine. Is it merely a coincidence that fetuses whose first nine months of life are characterized by extreme swings in available nutrition develop into adults who have metabolisms that facilitate survival in just that sort of nutritional environment? Scientists cannot yet answer this question, but Jones and I are both intrigued by the possibility.

SINGIN' IN THE EGG: THE PRENATAL DEVELOPMENT OF SENSORY PREFERENCES

There are several other ways in which environmental "information" can impact the development of a fetus besides via hormone receptors in the brain. For example, fetuses are now known to taste, smell, hear, and even see certain stimuli present in their uterine environments. Importantly, recent studies have demonstrated that such prenatal experiences can influence postnatal behaviors.* Among the noteworthy implications of these findings is the growing awareness among scientists that seemingly "instinctive" behaviors might, in fact, reflect the influence of these prenatal experiences. Not long ago, students of animal behavior assumed that behaviors characterizing *all* normal newborns in a species must be "instinctive," because they develop even in animals that have no chance to *learn* the behaviors. The finding that prenatal experiences influence postnatal behaviors suggests that such assumptions are unwarranted.

Consider the newborn mallard duckling's normal response to its mother's "assembly call"; this call attracts ducklings so that they approach their mother if she is stationary and follow her if she is moving. Early studies of this behavior revealed that ducklings that develop in an incubator— and so *never* hear the maternal assembly call during prenatal development—are nonetheless attracted to the call after they hatch. As a result, it is tempting to imagine that the attraction is "instinctive," or caused by genetic factors operating independently of the environment. The fact that ducklings' attraction to the call contributes to their survival (that is, that the attraction is evolutionarily adaptive) only strengthens this temptation.

Fortunately, some scholars of development refuse to be satisfied with such pseudo-explanations for early-appearing behaviors. One such scholar, a guiding light in the field of developmental psychobiology, is Gilbert Gottlieb. Gottlieb knew that if ducklings deprived of maternal stimulation respond to assembly calls nonetheless, then maternal stimulation must not be required for the development of the response. But more importantly, Gottlieb understood that this does *not* mean that the environment *in general* is unimportant in the development of the trait!* So, he began to look for other environmental factors involved in the trait's development. But what other factors could possibly explain this phenomenon? After all, we're now talking about an egg sitting undisturbed in an incubator in an empty room; surely any traits that develop in such a stimulus-free environment must be caused by genetic factors alone, right?

Gottlieb, like all good scientists, is incredibly observant. More than 30 years ago, he noticed that mallard duck embryos begin breathing and can vocalize a few days before hatching. And so it occurred to him: might ducklings' *own* vocalizations affect their later responsiveness to their mothers' assembly calls? This seemed unlikely at first glance, because there is almost no resemblance at all between the peeping of unhatched ducklings and the calls produced by mature mallard ducks. But, as we will see repeatedly in the next chapter, environmental factors that make essential contributions to the development of specific traits need not be *obviously* related to the traits they influence. In fact, Gottlieb found that depriving duck embryos of *their own* vocalizations actually does prevent them from responding normally to postnatal presentations of the assembly call; thus, the development of this response *depends* on the embryos having auditory experiences with their own vocalizations.

Both the traditional idea that "experience" begins only after birth and the related notion that behaviors present *at* birth must be experience-impervious "instincts" are obviously faulty. Even though *newly hatched* mallards in the wild are *always* attracted to their mothers' assembly calls, this attraction does *not* develop independently of experiential factors. Rather, in what is undoubtedly another example of an experience-expectant process, the development of this evolutionarily adaptive trait is powerfully influenced by the auditory experiences the embryonic duckling has inside its egg. As it happens, traditional understandings of "experience" and "instinct" are not just faulty; they actually jeopardize the search for genuine understanding. Labeling a behavior "instinctive" typically inhibits scientists' inclination to study its development, because there is little reason to investigate development if we "know" that a behavior is "instinctive" or "genetically determined."

"NOW, WHEN *I* WAS A FETUS . . ."

Human fetuses, too, are open to sensory experience; the research that yielded this discovery demonstrated that newborns' behaviors are affected by the sounds they experienced in utero. This research followed on the heels of the noteworthy revelation that if newborns are put into a situation where they can *choose* aspects of their experiences, they will choose to hear their mother's voice instead of the voice of another baby's mother. These exceptional insights have been won only recently because of the technical difficulties associated with learning anything at all about what is going on inside a newborn baby's head. Our contemporary understanding of newborns' preferences has resulted directly from the persistent and creative work of developmental psychologists intent on figuring out how to "get inside" babies' minds.

The methods developed by Anthony DeCasper to achieve this goal are particularly clever and useful. DeCasper knew that in order to learn about what goes on inside the head of an infant, the infant must be able to *do* something that can serve as an indicator of its "mental" activity. Unfortunately, there aren't that many things that newborn infants can do. But as long as there is *some* behavior that the newborn can control, hope lives on. One of the things that newborns do fairly well right after birth is suck, so, as unlikely as it seems at first, sucking can be used as a "window" through which we can peer into a small corner of a baby's mind.

To take advantage of infant sucking, DeCasper and William P. Fifer wired a pacifier nipple to a device that allowed them to record every suck produced by infants in their study. Initially, they knew that when newborns suck on a pacifier, they do not just suck away continuously. Instead, they typically release a burst of several discrete sucks that occur at a rate of about one per second; each burst is typically followed by a longer "rest" period, in which the infant does not suck at all for several seconds. So, after fitting infants with both headphones and the specially wired pacifier, DeCasper and Fifer allowed the babies to suck in silence for a while. During this period, the lengths of each infant's rest periods were measured. While the rest periods taken by an individual baby vary in duration, each infant has an *average* length that characterizes its rest periods. So, DeCasper and Fifer then played the voice of the infant's mother (through the headphones) *only when the baby altered the durations of the rest periods taken between sucking bursts.* Specifically, some newborns heard their mother's voice only if they began taking longer-than-average rest periods; if they began taking shorter-than-average rest periods, they heard the voice of a *different* baby's mother (reciting, by the way, the same poem recited by the baby's mother). Other newborns were required to generate shorter rest periods to hear their own mother and longer rest periods to hear a different baby's mother. And, as I have noted, these babies (less than *three days old!*) changed their sucking patterns in order to hear their own mother's voice. How can we explain this phenomenon?

There are two or three possible explanations for the finding that new-borns prefer their mother's voice to that of an unknown woman. Although the newborns in DeCasper and Fifer's study had no more than 12 hours of postnatal contact with their mothers before testing, it remains possible that this much exposure is enough to account for the observed preference. DeCasper and Fifer noted that newborns might be able to learn so quickly that "limited postnatal experience with the mother results in preference for her voice." Alternatively, since third-trimester fetuses respond to sound and so can probably hear, it might be that *prenatal* exposure to the mother's voice accounts for the observed preference. DeCasper and Fifer ended their 1980 article noting that "although the significance and nature of intrauterine auditory experience in humans is not known, perceptual pref-erences . . . [of some nonhuman] infants are profoundly affected by audi-tory experience before birth." One final option—that the preference is "innate," "instinctive," or somehow directly caused by the genes—was not even entertained by DeCasper and Fifer; apparently, these scientists understood that this sort of explanation is, in fact, no explanation at all.

Within one year, DeCasper and Fifer's hunch about the importance of prenatal experience found important support: French researchers managed to record sounds from *inside* the uteruses of pregnant women about to give birth. What they discovered was that the sound of the maternal voice *is* audible inside a uterus distended by nine months of fetal growth.* This finding encouraged DeCasper and Melanie J. Spence to devise a study that would more directly test the hypothesis that fetuses' auditory experiences affect their preferences and behaviors as newborns. They wrote:

> The prenatal experience hypothesis implies that newborns prefer their own mothers' voices . . . because of prenatal experience with her voice-specific cues. This implication, however, cannot be directly tested for obvious ethical and practical reasons [such a test would require asking pregnant women to refrain from speaking during the whole of their pregnancy!]. The hypothesis also implies that newborns will prefer the acoustic properties of a particular speech passage if their mothers repeatedly recite that passage while they are pregnant.

Thus, DeCasper and Spence asked mothers in their seventh month of pregnancy to recite a children's story aloud twice every day for the last six weeks of their pregnancies.

Simplifying somewhat (since three stories were actually used), this is what DeCasper and Spence did. First, they asked half of the pregnant mothers in their study to recite *The Cat in the Hat* daily, while the other half recited *The King, the Mice, and the Cheese*; these passages are of similar length and complexity, but they contain different words and rhythmic characteristics. Then, two days after the babies were born, they were tested using the wired pacifier-headphone apparatus described above. Half of the infants were required to generate longer-than-average rest periods between sucking bursts in order to hear a recording of a mother reciting *The Cat in the Hat*; if these infants generated shorter-than-average rest periods, they heard a recording of the same mother reciting *The King, the Mice, and the Cheese*. The other half of the infants had to generate *shorter* rest periods to hear *The Cat in the Hat* or *longer* rest periods to hear *The King, the Mice, and the Cheese*. Finally—as if the study were not already complicated enough— half of all tested babies heard a recording of their *own* mother reciting *The Cat* or *The King* (whichever the baby "chose"), while the other half heard a recording of a *different* baby's mother reciting the "chosen" passage. This last variation allowed the researchers to address the impact of each mother's specific vocal characteristics on her baby's preferences.

It turned out that babies did what they needed to do to hear whichever passage their mother had recited aloud throughout the final weeks of her pregnancy. *Moreover, it turned out that it didn't even matter if the woman reciting the story was the baby's own mother or another baby's mother!* If a mother had recited *The Cat in the Hat* during her pregnancy, her newborn baby changed its sucking pattern in order to hear *The Cat in the Hat*, regardless of whether the available recording contained the mother's voice or the voice of a different baby's mother. As DeCasper and Spence concluded,

> The only experimental variable that can systematically account for these findings is whether the infants' mothers had recited the target story while pregnant . . . The most reasonable conclusion is that the target stories were . . . preferred, because the infants had heard them before birth.

These results have since been widely reported in textbooks and in the popular media, as they deserve to be; the implications of the fact that fetuses are affected by the sounds they hear could be extremely important.* But it turns out that some of the more significant implications of this research are actually not about fetal sensitivity to sounds per se at all. Instead, this research can help us understand some general points about the role of the environment—and the experiences that it affords—in the development of our traits. First, these results underscore the importance of recognizing that "experience" begins at conception, not at birth. Second, they show the importance of acknowledging that all development occurs in *some* environment, even if it occurs in an environment that we have not previously recognized as such (for example, a uterus). Finally, it supports the idea that prenatal experiences can affect both the fetus's brain and the newborn's subsequent behavior. While genetic factors impact the development of *all* traits—without genes, fetuses would have no ears or brains, rendering DeCasper and Spence's results impossible—the ever-present environment, too, has the ability to impact development in subtle, often unexpected, but probably very important ways.

RIGHT BRAIN, LEFT BRAIN

One of the great developmental thinkers of our time, Gerald Turkewitz, has speculated recently about the importance of environmental contributions to the prenatal development of *cerebral lateralization*. Discovered through the Nobel Prize–winning research of Roger Sperry and Michael Gazzaniga in the 1960s, "cerebral lateralization" refers to the well-publicized finding that, in most adults, the right and left sides of the brain operate somewhat independently and specialize in processing different kinds of information. In particular, the left side of the brain appears to specialize in processing linguistic stimuli and in performing logical, analytical tasks such as reasoning. The right side of the brain, in contrast, appears to be better at processing spatial information, or, more specifically, at performing holistic pattern detection tasks like recognizing faces; the right side of the brain also seems to specialize in processing musical information, at least in nonmusicians.*

Cerebral lateralization (including specialization of function) is present at birth, at least to some degree. As proof of this fact, Turkewitz cites a study by M. Hammer in which newborns were presented with either speech sounds or nonspeech noise. Hammer found that newborns preferentially turned their eyes toward the right when they heard a sample of speech, but toward the left when they heard nonspeech sounds. These data, Turkewitz writes, "support the view that 2- to 3-day-old infants . . . [have] a right-ear (and, possibly, a left-hemisphere) bias for processing speech and left-ear (right-hemisphere) bias for processing noise."

In response to the finding that even newborns' cerebral hemispheres have different functions, many observers concluded that lateralization must somehow be specified in the human genome; these observers could see no other way to explain the fact that asymmetry characterizes almost *all* human brains, even at birth. By this point in the book, however, it might be clear how inappropriate such conclusions really are. Turkewitz writes:

> It has been assumed that knowing that a particular function was present at birth is equivalent to understanding its source as well as implying a sort of developmental invariance. Presence of capacities at birth has frequently led to these capacities being described as hard-wired or genetically determined. Such thinking is by no means new and in fact has been cogently argued against in the past.

He then speculates that cerebral lateralization could actually have its origins in the gross particulars of a fetus's developmental circumstances.

Turkewitz initially points out that characteristics of the uterus change as pregnancy proceeds; as a consequence of these changes, the sounds that fetuses hear in utero change in the months before birth. In particular, the sounds audible to fetuses earlier in pregnancy consist primarily of noises generated by the mother's body (for example, her heartbeat, the sound of blood coursing through her veins and arteries, the sounds of her digestion, and so on). In contrast, later in pregnancy, the sounds audible to fetuses include speech produced by the mother and by other people in the mother's environment. Turkewitz then notes that because the fetus's right and left cerebral hemispheres begin to develop at different times and subsequently develop at different rates, the types of information they ultimately specialize in processing could differ as a function of the sounds they were *initially* required to process.

Specifically, Turkewitz hypothesizes that the slightly earlier developing right hemisphere might take on the task of processing the type of information that is available in the intrauterine environment early in gestation, namely nonspeech information. Later on, when speech information has become more audible and the left hemisphere has become more developed, this sort of information might be processed by this hemisphere. The idea is that at this point in time, the left hemisphere might be the only hemisphere *available* for such processing, since the nonspeech stimuli initially processed by the right hemisphere continue to be present and to monopolize the processing resources of the right hemisphere. In this way, Turkewitz writes that linguistic information

> could come to be processed by the left hemisphere both by default and because of its emergence as the now more advanced structure. It is important to note that . . . relative differences between the hemispheres, even if small in absolute terms, could be sufficient to produce important structural and functional differences.

Thus, Turkewitz thinks that the combination of the different rates of right and left hemisphere development, coupled with the changes that occur in "the nature of the fetal acoustic environment, [could together] help to shape hemispheric specialization." Turkewitz realizes that his proposal is highly speculative, of course, but I find it fascinating nonetheless.

Even more remarkable is Turkewitz's suggestion that lateralization could be further influenced by the posture and limited mobility that characterize normal fetuses in their final 10 weeks in utero. As fetuses become larger, gross changes in position become increasingly difficult. By the end of pregnancy, then, fetuses are normally suspended in utero head-down and facing toward their mother's right side. In this position, their left ear is closest to their mother's spine and their right ear is closest to their mother's abdominal wall. This arrangement, according to Turkewitz,

> would be likely to result in differential exposure of the two ears to . . .
> maternal speech propagated [down the spine] by bone conduction.
> This position would also result in differential exposure of the two ears
> to externally generated sound. . . . Fetal posture would accentuate
> asymmetric differences in the nature of the acoustic environment and
> could contribute to the development of the hemispheric specialization
> under discussion.

While data bearing on this hypothesis are not yet available, it seems reasonable to me at first glance, and probably warrants further examination.

Turkewitz's speculations must be considered with caution, since they are, after all, speculations; still, his creative ideas stretch our conceptions of how traits develop and thus serve to exercise our minds in ways that are extremely stimulating. For the moment, the most important lesson to take from them—and from DeCasper, Gottlieb, and Jones's research, as well—is that valuable insights can arise from thinking deeply about what actually occurs during development. If we assume that traits that are present at birth must be genetically determined, we will be blinded to the nongenetic factors that contribute to the prenatal development of these traits—even if we are talking about "biological" traits that develop universally in normal human beings. And, in ignorance of the influence of these prenatal factors, we could miss out on relatively inexpensive or easy ways to affect those traits in desirable ways.

9

CHICKEN SHOES AND MONKEY FOODS
The Not-So-Subtle Effects of Some Very Subtle Postnatal Experiences

During the past two years I have had to spend periods of several weeks on a remote island in comparative isolation. In these conditions I noticed that my beard growth diminished, but the day before I was due to leave the island it increased again, to reach unusually high rates during the first day or two on the mainland. Intrigued by these initial observations, I have carried out a more detailed study and have come to the conclusion that the stimulus for increased beard growth is related to the resumption of sexual activity.

So begins one of the most unusual articles I have ever encountered in the scientific literature, a body of writing sometimes so tedious as to put even the most curious of minds to sleep within a matter of pages. But buried within this expanse of articles are some truly strange, surprising, and, in the best of cases, enlightening reports of phenomena that force us to consider our unexamined assumptions about how we work. A close look at some of these phenomena—including the effect of sexual activity on beard growth—will make it clear that our environments and genes, both, continue to influence our traits through our entire lives.

Once an infant is born, whole new categories of macroenvironmental experience become available to affect trait development. For example, as soon as they are born, babies have their first experiences with other living things, with objects, and with patterned light (as opposed to the very dim, diffuse light that fetuses might see). The effects of these experiences are, in some cases, not surprising. In other cases, though, the effects were unanticipated before they were discovered. Just like the impact of a mallard duck embryo's own prenatal vocalizations on its later preference for its mother's call, some of the *postnatal* experiences that affect development do so in decidedly nonobvious ways.

CHICKEN SHOES AND MONKEY FOODS

These experiences can have profound, if unexpected, effects. Consider, for example, the tendency of two-day-old chicks to approach, pick up, and then eat mealworms present in their environment. In a clever but low-tech experiment, J. Wallman examined the effect on this response of preventing newborn chicks from seeing their toes. Because a chick's toes resemble mealworms in size, color, and segmentation pattern, Wallman hypothesized that visual experience with their own toes "perceptually prepares" chicks for encounters with mealworms (even though mealworms and chicks' toes differ in a variety of ways, including their movement patterns).

To test his hypothesis, Wallman restricted the visual experiences of chicks in a minimal way. Shortly after they hatched in a dark incubator that prevented them from seeing their feet, he fitted several chicks with cloth "shoes" that permitted walking but not toe inspection; several other chicks were treated identically, except they were left shoeless. After two subsequent days of otherwise normal experiences in the world, all of the chicks were observed for 5 minutes while in the presence of a mealworm. During this period, chicks that had never laid eyes on their own toes were significantly less likely to pick up or eat the mealworm than the "barefoot" chicks; instead, chicks with covered feet mostly just stared at the mealworm, one eye at a time.* Thus, the simple lack of visual experience with their own toes was enough to interfere with a response that—because it appears in *all* chicks hatched in normal environments—had looked "innate" prior to Wallman's research. Wallman wrote:

> The effect of early visual experience is not simply to facilitate perception of those objects present in the early environment, but also to facilitate perception of other objects that have some attributes in common with the objects seen. From this point of view, everything the animal sees might influence . . . its perceptual development.

Such effects are also found in mammals, and they are not limited to visual experience with things that vaguely resemble later misperceived stimuli. For example, consider the development of snake phobias in primates. Since the nineteenth century, it has been known that wild-reared adult chimpanzees are intensely afraid of snakes; this observation has now been extended to include many primate species. In a remarkable study of the nonobvious effects of experience, N. Masataka demonstrated that squirrel monkeys can develop a fear of snakes based not on exposure to snakes per se—or even to stimuli that look like snakes—but rather on exposure to insects. Masataka's study involved looking at how 24 monkeys who were at least 10 years of age responded to real, toy, and model snakes. Eight of the monkeys tested had been born and raised in the wild, and eight had been born and raised in a laboratory where they had had no contact with any live animals and where they had been fed a diet consisting only of fruits and monkey chow. The final eight monkeys tested were exactly like the other laboratory-reared monkeys, except that four and a half years before the start of testing, the experimenters began to add one grasshopper or one cricket each day to what had previously been a strictly fruit-and-monkey-chow diet.

When these various monkeys were finally exposed—sequentially, but in different orders—to a live boa constrictor snake, a nonmoving lifelike model of a snake, a rubber toy snake, black and yellow rubber electrical cords, and a variety of neutral wooden blocks, their responses varied as a function of their prior experiences. Specifically, the wild-reared monkeys and the laboratory-reared, insect-fed monkeys behaved as if they were afraid of the real snakes, the model snakes, and the toy snakes (but not of the electrical cords or neutral objects).* In contrast, the fruit-and-monkey-chow-only monkeys behaved fearlessly in the presence of all the test stimuli. Thus, simple exposure to insects led laboratory-reared monkeys to fear snakes just as the wild-reared monkeys did, even though the life experiences of these two groups were extremely different. And, although the *only* difference between the laboratory-reared groups was their exposure to insects, the insect-fed monkeys responded to snakes very differently from the fruit-fed monkeys.

Exactly *how* a snake phobia can be instilled via insect exposure is not apparent from this study, but Masataka speculated that the insect-fed monkeys' "wider perceptual experience with living beings [might have] . . . enabled them to distinguish between edible and dangerous animals." But regardless of *how* a monkey's experiences contribute to the development of a normal snake phobia, the point remains the same: the snake phobia that characterizes *all normal* squirrel monkeys *requires* certain postnatal experiences for its development. Furthermore, the development of this trait depends on an experience that bears little obvious relationship to the trait it affects.

This sort of finding has significant implications for researchers studying trait origins. A common method of judging a behavior to be "innate" or "instinctive" involves seeing if the behavior develops even when animals are deprived of experiences thought to be necessary for the behavior's development. But if, as Masataka's and Wallman's results suggest, it is nearly impossible to guess in advance which experiences really are the important ones, then this method can effectively create an *illusion* that a trait's development does not require any experience at all. For example, a researcher interested in determining if male rats' sexual behaviors are innate might proceed by depriving male rat pups of exposure to female peers and to parental displays of sexual behavior. If the pups then develop into adults that exhibit normal sexual behavior anyway, the researcher might be lulled into the conclusion that these behaviors are innate, instinctive, or genetically determined (depending on the researcher's specific theoretical biases).

But drawing such a conclusion would be a mistake, as indicated by data on the development of real male rats' sexual behaviors. It turns out that mother rats normally stimulate their newborn pups for the first few weeks of their lives by licking them in the area between the anus and the urinary tract opening; this stimulation supports the pups' waste elimination. Typically, male offspring receive more licking than female offspring, due to the fact that the glands that secrete the chemical eliciting this maternal licking produce more of the chemical in male than in female pups. Celia Moore has shown that this licking affects the sexual behavior of the pups once they mature. Specifically, experimentally understimulated males exhibit deficient male sexual behavior, and they have fewer neurons in spinal cord areas that normally support such behaviors. Thus, both behaviorally and neurologically, the development of normal male sexual behavior in rats does require the rats to have certain specific experiences, even if the experiences in this case are not the ones we might have *thought* beforehand would be the important ones.

The effects of neonatal anal-genital licking have no direct analogs in human development (at least, one can hope!); nevertheless, there are important general lessons to be taken from these sorts of phenomena. First, *experiences with the macroenvironment can directly affect the structure and function of the nervous system; such experiences, then, can potentially influence any behavior or mental activity.* Second, *species-typical traits*—including those that *look* "innate"—*can depend for their development on experiences that are universally encountered by individuals developing in normal environments;* all normal chicks eat mealworms, not because this behavior is genetically determined, but because chicks that develop in normal environments *always* see their toes shortly after hatching. Finally, *developmental experiences can affect mature behaviors in nonobvious ways;* few of us would have guessed that early exposure to such un-snakelike animals as insects could play a role in the later development of a normal snake fear. Taken together, these observations weigh powerfully against credulously accepting the facile conclusion that a given trait can develop without experiential input. The fact of the matter is that trait development is extremely complicated, including roles for both microenvironmental factors and for macroenvironmental factors that bear little resemblance to the experiences we often think of as contributing to learned behavior.

ENGAGED WITH OUR ENVIRONMENTS, 'TIL DEATH DO US PART

It is part of the public "intuition" that many of our adult traits are established early in life and are somehow rendered unchangeable by later experiences. In fact, early experiences *are* extremely important in the development of many of our traits; I have presented data that are consistent with this view.* Nevertheless, the idea that our characteristics are *fixed* by the time we reach adulthood—let alone the popular idea that they are permanently established by the time we enter elementary school—is not consistent with what scientists now know about how genetic and nongenetic factors interact throughout our lives.* Early experiences certainly contribute significantly to trait development, but even if they turn out to be *more* important, in some respects, than later experiences, this does not mean that later experiences do not appreciably affect our traits.

Even as many of us believe in the permanent effects of early experience, many still believe that the more we age, the more our psychological characteristics reflect our experiences. The latter belief no doubt reflects the same sort of reckoning that leads us to misattribute newborns' traits to their genes alone: by the time we are old, we have had many experiences, and we are sometimes acutely aware of how these experiences have affected our personalities. Thus, many people believe that our genes do most of their work on our bodies and minds when we are young and that when we are adults, our genes are largely quiet; having done the work of "building" us, they just sit back and let us go about the business of living.

But in fact, it doesn't work like this at all. Through our entire lives, our genes continue to do what genes do: make proteins. And through our entire lives, our macro- and microenvironments continue to do what the environment does: provide the nongenetic factors that co-act with genes to produce traits. Genetic and nongenetic factors influence the biological and psychological characteristics of *mature* animals just as surely as they influence the characteristics of juveniles.

On Hormones and Pheromones

Some traits remain dynamic across the life span because of the ability of the environment to affect hormones that can influence genetic activity. The ramifications of this arrangement are just amazing: if macroenvironmental events can affect hormone levels, and hormones can control genes, then the events of our daily lives can effectively control our genes. But can macroenvironmental events really have significant effects on hormone levels? Absolutely.

When male doves of a particular species want to mate with a female, they perform specific behaviors in front of her, behaviors that depend on the presence of testosterone in their bodies. But the male's testosterone level itself is affected by a variety of factors, including day length; testosterone levels increase as the days get longer in the springtime. So, how can something like day length influence hormone levels? Actually, it's fairly easy: there are neurons whose axons carry information about the presence of daylight from the eyes back to the part of the brain (the hypothalamus) that controls the pituitary gland's secretion of hormones; the more hours of daylight, the more testosterone. In addition, hormone levels can be affected by *specific* auditory and visual stimulation: a male secretes more testosterone when in the presence of a female dove, and a female secretes more estrogen when she sees and hears a male dove courting her. Moreover, as Michel and Moore note, "stimuli provided by a mate sometimes even affect the hormonal status of an individual indirectly," through feedback from the behaviors the individual performs upon encountering the mate. If a similar feedback process operated in people, it would be as if some of *my* behaviors—for instance, if I unconsciously comb my fingers through my hair whenever I see a beautiful woman—could themselves induce my cells to produce more testosterone. But enough about hypothetical processes: can environmental stimuli actually alter *human* hormone levels?

For the past 25 years, airborne molecules known as *pheromones* have been understood to influence the behavior of several mammalian species.* These molecules are released into the macroenvironment by individual animals and, after they are detected in the nose of another animal of the same species, they influence the behavior and/or physiology of the recipient animal. But while proof of the existence of pheromones in animals like hamsters, deer, and elephants has been available for some time, evidence of the existence of human pheromones was lacking until 1998.

Women know that living with other women can lead to synchronization of their menstrual cycles. But it was not until Kathleen Stern and Martha McClintock began to look for pheromonal control of this phenomenon that it was finally explained and the existence of human pheromones was definitively demonstrated. In their study, Stern and McClintock asked nine women in different phases of their menstrual cycles to bathe without perfumed products and then to wear cotton pads in their armpits for eight hours a day. The pads provided by these donors were subsequently treated with alcohol and later wiped under the noses of 20 healthy, young female volunteers on every day of their menstrual cycles (these volunteers were not on birth control pills). Importantly, the volunteers—who had no idea what the study was about or where the pads had come from—never reported conscious awareness of an odor other than the smell of the alcohol with which the pads had been treated (this is important because pheromones are usually defined as *odorless* molecules that are nonetheless detected in the nose).

The results of the study were clear-cut. Women who were exposed to control pads containing *only* alcohol experienced no changes in the lengths of their menstrual cycles. In contrast, most of the volunteers exposed to pads worn by *ovulating* donors experienced lengthened cycles, whereas most of the volunteers exposed to pads from donors who were at earlier points in their cycles experienced shortened cycles. Amazingly, exposure to these odorless—and so consciously undetectable—stimuli affected the amount of time that elapsed before the volunteers' *very next menstruation*, increasing it or decreasing it, depending on the pheromone they were exposed to. Measurements of the volunteers' hormonal statuses were equally revealing: exposure to pheromones generated by ovulating donors delayed the volunteers' normal surge of ovulation-inducing hormone, whereas exposure to pheromones generated by donors at earlier points in their cycles sped up this surge. Stern and McClintock concluded that their data provide "definitive evidence of human pheromones," "demonstrate that humans have the potential to communicate pheromonally," and identify "a potential pheromonal mechanism for menstrual synchrony." This, then, is direct evidence that macroenvironmental stimuli can significantly impact human hormone levels. In addition, it opens up the possibility that environmental events can influence gene activity through hormone activity, since various hormones are known to be able to "turn genes on and off."

The words of the anonymous island-bound scientist* whose tale begins this chapter tell a similar story, although in this case, the hormone response was not pheromone-induced. Every day, "Anonymous" weighed the shavings recovered from the head of his electric razor and recorded data on his activity levels, sleep, libido, and intercourse. As noted above, a study of the correlations among these factors revealed "a most marked increase in beard growth" on "the day of return to the mainland and the initial resumption of sexual activity." Ultimately, Anonymous traced the likely cause of this effect to the ability of intercourse to increase men's testosterone secretion. He did this by measuring his beard growth in response to oral ingestion of both placebos and hormones; as expected, ingestion of any of the androgens he tried (testosterone, androsterone, and methyl testosterone) led to increased beard growth similar in magnitude to the increases he saw after resuming sexual activity. Thus, these data hint at a pathway by which environmental events can alter bodily traits, namely via changes in levels of steroid hormones known to influence genetic activity.

Anonymous also reported changes in his beard growth that seemed to occur as a consequence of significantly subtler events in his life. For example, he wrote, "even the presence of particular female company in the absence of intercourse, after a period of separation, usually caused an obvious increase in beard growth." Moreover, he even reported a marked *anticipatory* effect, such that his beard grew significantly more on Fridays— *before* he departed, ever hopeful, for a sexual encounter on the mainland—than on any of the other lonely days of the week. This is truly remarkable insofar as it suggests that *purely psychological events occurring in the absence of physical stimuli* might be able to affect the activity of our genes (by way of the mediating influence of our hormones).

Of course, hormones like estrogen and testosterone are not the only steroid hormones in our bodies that can mediate between environmental events and genetic activity. When we are under stress, the so-called stress hormones are released by our adrenal glands from their position above our kidneys. These hormones, like the sex hormones, can diffuse into cell nuclei where they can bind with hormone receptors; thereafter, they too can bind (as a steroid-receptor complex) to DNA, effectively turning genes "on" or "off," or regulating the rate at which genes are decoded. It has been known for some time that prolonged psychological stress can affect how well an adult's immune system responds to agents of infectious illness like viruses or bacteria; it is likely that one way in which stress contributes to illness is via the influence of hormones released in response to environmental stressors.

Findings like these are exceptionally important because they illustrate how experiential factors such as day length, mate presence and behavior, and the presence of psychological stressors can affect our bodies and behaviors by affecting hormone levels in our blood streams. Many of the psychological and biological traits we have as adults remain "open" to environmental influence by way of this mechanism; our experiences as adults can affect our physiology, our genetic activity, and our brains, thereby influencing our behavioral, cognitive, and affective traits. But the story does not end here: recent empirical evidence suggests that macroenvironmental events can *directly* affect genetic activity as well, even without the mediating contributions of hormones.

Immediate-Early Genes

Some studies demonstrating the existence of direct environmental effects on genes have examined the activity of so-called *immediate-early genes* (IEGs). IEGs are genes that can be activated directly by signals arising in the macroenvironment. K. M. Rosen and colleagues have written that "there is increasing evidence that [IEGs] play an active role in converting brief environmental input into lasting changes in cellular function. [These genes] are rapidly activated in neurons" in response to signals arising outside of the cells. According to Michel and Moore, "IEGs are therefore an important link in the cellular machinery that can convert brief environmental input into enduring cellular changes, thereby contributing both to the construction of the developing nervous system and to its plasticity [that is, adaptability] throughout life."

One of the early studies of IEG activity was conducted to explore how mammals' circadian rhythms are affected by exposure to light. The phrase "circadian rhythm" refers to the fact that many of our biological activities occur in regular cycles that are close to 24 hours in length. One way to become acutely aware of these rhythms is to interfere with them; this happens after flights across time zones, which produce the phenomenon known as "jet lag." For example, on days when I fly from New York to Los Angeles, my day is 27 hours long (because California is three hours behind New York); as a result, my circadian cycles—which at that point are adjusted to New York time—start over again before my day really ends. Consequently, in the days following such trips, I will invariably be wide-awake at 5 A.M. in Los Angeles, because it will feel to me like it is already 8 A.M. (since it *is* 8 A.M. in New York, where my circadian "clock" was set). The question is, how come one week later I will have no trouble being asleep in Los Angeles at 5 A.M.? How do my circadian rhythms readjust to "local time"? It turns out that circadian rhythms are synchronized with local time by way of small daily shifts in the brain's "biological clock"; this is why it takes a few days to recover completely from jet lag. Furthermore, these shifts are known to result from exposure to light during what *feels* like nighttime (that is, the shifts are produced in response to the *light* I see in Los Angeles at 5 P.M. when it is already 8 P.M. and dark in New York). But the question remains: How exactly can light exert this sort of effect on our brains?

B. Rusak and his colleagues studied this phenomenon in hamsters, because hamsters, like people, have circadian rhythms that adjust to local day-night cycles. First, Rusak exposed one group of adult hamsters to a normal light-dark cycle, and another group to 30 minutes of light turned on in the middle of the night. Then, just after each group was exposed to the light, Rusak examined an area of their brains that normally receives information from light-detecting neurons in the eyes. The results suggested that under normal circumstances—that is, 12 hours of darkness followed by 12 hours of light—no immediate-early genes were activated. In contrast, when hamsters were examined immediately after having been exposed to a half hour of light in the middle of the night, rapid physiological changes had occurred in the particular brain area examined. Specifically, brief exposure to light in the middle of the night led to dramatically increased quantities of particular types of RNA in this brain area, indicating that the immediate-early genes that serve as templates for these RNAs had been activated—"turned on"—as a result of the visual stimulation. Rusak concluded that "cells in this portion of the [hamster brain] undergo alterations in gene expression in response to . . . illumination, but only at times in the circadian cycle when light is [not ordinarily experienced]." Thus, a simple experience like seeing light when darkness is "expected"—which happens regularly after flights across time zones—can actually turn specific genes "on" in adult mammals' brains.*

Data on the effects of more complex "psychological" stimuli on genetic activity similarly indicate that genes and environments co-act throughout our lives. In an elegant study of canaries and zebra finches, C. V. Mello, D. S. Vicario, and D. F. Clayton examined whether or not IEGs in cells from certain brain areas might be activated in response to birdsong. Adult songbirds such as canaries and zebra finches use song to communicate with others of their species about important things like reproduction and territoriality; in response to hearing the song of another bird of their species, these birds change their own social and vocal behaviors. Thus, birdsong, while macroenvironmental and quite complex, is a natural auditory stimulus that is very important to these birds.

Mello and colleagues first isolated each of their birds from other birds for an entire day. Then, once a minute for 45 minutes, they repeatedly exposed each bird to a short tape-recording of a particular test sound. Specifically, some of the birds heard a song that is characteristic of their species, others heard a song that is characteristic of the *other* species (that is, canaries heard a zebra finch song and zebra finches heard a canary song), and still others heard tone bursts (a non–song sound that had the same range of frequencies as birdsong); a fourth group heard no sound at all. Subsequently, the researchers looked for evidence of IEG activation in brain areas connected with primary sound-processing areas. Sure enough, they discovered that "the brains of songbirds respond to song playbacks [but not to tone bursts or to silence] with a large and rapid increase" in the production of RNA associated with particular IEGs. Moreover, the songs produced by a bird's own species "elicited the highest response, whereas other classes of auditory stimuli were less effective or ineffective." They concluded that these experiments provided direct evidence of "a role for genomic responses in neural processes linked to song pattern recognition."

Experiences directly affect *human* gene activity as well, even in mature adults. While the exact mechanisms by which stress interferes with our immune systems is still being worked out, recent data indicate that psychological stress can turn certain genes off directly, and not only via the mediating influence of stress hormones. Specifically, R. Glaser and colleagues studied the white blood cells (which are part of the immune system) of a group of first-year medical students, both when they were unstressed and in the midst of a stressful three-day period of academic examinations. Glaser and colleagues reported that when students were stressed, they had white blood cells that contained less of the RNA that codes for a protein receptor molecule involved in the recognition of foreign substances like viruses and bacteria. This means that stressed students' white blood cells were less able to perform their immune functions; the reduced levels of RNA suggest that certain genes in the white blood cells—genes that, in the absence of stress, code for the needed protein receptor—had been "turned off." Glaser and colleagues wrote, "while there are ample data demonstrating stress-associated decrements in the immune response in humans and animals, [our] data provide the first evidence that this interaction may be observed at the level of gene expression." Thus, our daily experience of stress directly impacts the activity of our genes, changing our very constitution.

Environmental Information (and the Supposed Primacy of Genes)

On encountering data like these, some of my more skeptical colleagues have argued that, to them, it looks like environmental factors merely *support* normal development or, at best, "trigger" gene activities. Proponents of this view typically see the genes as containing coded "information." In contrast, they see the environment as merely providing the genes with raw materials needed to carry out genetic instructions; the environment itself is not recognized as a source of information. But it turns out that the validity of this argument depends on how we define the word "information." Consider, for example, the following definition, originally worked out in the 1940s by theorists studying communication: "information" is that which produces only one of at least two possible states in a "receiver" (in this case, an animal) because of the influence of a "sender" (in this case, genetic or nongenetic factors).* Given this widely accepted definition, the environment, *no less than the genes*, is a true source of information for an animal.

We can see this by reconsidering the findings of Mello and colleagues and Stern and McClintock. Recall that canary bird song *but not zebra finch bird song* induced gene activity in canary brains, and that zebra finch bird song *but not canary bird song* induced gene activity in zebra finch brains. Thus, different patterns of stimulation had different effects on different birds, including altering the activity of their genes, altering the presence of significant molecules in their cells, and even altering their behavior. Given the different consequences of exposure to these different songs, it would be unreasonable to maintain that genetic factors provide organisms with developmental information in a way that environmental factors do not; variations in either type of factor can cause variations in gene activities, cell structures,* and behaviors. Similarly, the finding that different pheromones had diametrically different effects on the women studied by Stern and McClintock means that it is a mistake to maintain that the environment does not "inform" developing systems. The hormones measured by these researchers were influenced in precise ways by the tested pheromones; the hormonal changes can be understood to have been responses to "information" carried by these pheromones.

The results of these studies also undermine our "intuitive" sense that trait determination begins with our genes. Even those who realize that genes are not exclusively responsible for giving organisms developmental information still sometimes maintain that the role of genes in trait development is more important than the environment's role because genes operate *prior to* the operation of environmental factors. However, the ability of macroenvironmental events (for example, psychological stressors and pheromones) and microenvironmental events (for example, hormones) to influence gene activity cripples this argument. Instead, these abilities encourage us to imagine developmental pathways as nonlinear, complex "domino trails" *that can circle back on themselves.* In fact, it is a mistake to think that genes are necessarily the first link in causal chains of biological events. In real organisms, cascades of biological events are often initiated by *non*genetic factors. From the moment of conception, environmental factors *and* genetic factors are in an ongoing "dialogue" with one another about building a person. Each of these sets of factors brings its own necessary information to this conversation.

Appearances to the contrary, our traits remain open to change throughout our lives, depending on both the state of our genes and the state of the nongenetic factors that surround them. Even traits that seem "fixed," such as hair color, are not unchangeable (my hair is turning gray as we speak!). If some traits *seem* fixed, it is only because genetic and nongenetic factors co-act to replace with identical cells, molecules, and pieces of molecules, the cells, molecules, and pieces of molecules that are damaged as we age or undergo certain experiences; this analysis holds for all of our cells and molecules, including those constituting our brains (which is why it applies to psychological as well as biological traits). Thus, the dynamic nature of our traits is sometimes invisible to us; this changeability becomes apparent only in the face of genetic or nongenetic *change*. Recent research has shown how some unusual changes in experience can lead to significant changes in the structures of our brains, even through adulthood.

PLASTIC BRAINS

Some of the most exciting brain research conducted over the past decade has demonstrated a remarkable sort of flexible responsiveness—called *plasticity* by neuroscientists—that shows how the structure and function of a mature brain remains open to change across the life span. As I noted above, studies of people deaf or blind from an early age have found that auditory or visual cortex can be recruited to process visual or auditory information, respectively. But in these studies, the participants were immature when they first lost sensation in one of their sensory modes. As surprising as the plasticity of these people's brains turned out to be, the fact that such plasticity was found in people who lost a sense when they were young did not surprise developmentalists. Remember, every cell in young human embryos can become *any* of the different cell types that make up a human body, and undifferentiated neuroblasts—which, you will recall, normally develop into neurons—are extremely plastic early in life. Thus, developmentalists have come to *expect* extraordinary plasticity from immature animals' brains. In contrast, the *adult* brain has typically been thought to be less responsive in the face of environmental challenges. Nevertheless, recent research has called this view into question. Flexible responsiveness now appears to characterize brain recovery after damage in adulthood as surely as it characterizes brain development. This new understanding has been borne out in studies of rodents as well as in studies of our closest animal relatives, the primates.

In parts of the normal adult human brain, sensory "maps" of the body exist, such that the *perception* of someone touching your palm depends on activity in a brain area that is close to the brain area where you process information about someone touching your wrist; since your palm and wrist are close to each other, the brain areas that receive information from these body parts are close to each other as well. This arrangement is maplike because distances across the brain represent distances across the skin just as dots symbolizing Boston and New York on a map represent the geographical distance between these cities. The same story holds for brain *control* of behavior: the brain areas that control movement of neighboring body parts are closer to one another than are the brain areas that control movement of body parts that are farther from one another. Questions about the plasticity of these maps require looking at whether or not they change as a result of experiences in adulthood.

Studies of such maps in the brains of primates who have had a portion of a finger experimentally amputated as adults* have revealed that cortical maps remain plastic into adulthood. Specifically, in response to this experience, monkeys' brains have been shown to reorganize themselves, so that brain areas previously devoted to sensing and controlling the now-lost digit change their function, and instead begin sensing and controlling neighboring digits. Nonexperimental studies of human amputees, likewise, have contributed to the belief that amputation "causes the grand-scale functional reorganization of maps on the surface of the brain representing body sensations and movements." These results support the idea that "any given skin surface can be represented by many alternative functional maps at different times of life" and that the "basic features of . . . cortical maps . . . are *dynamically* maintained," meaning that they are not determined by early experience and subsequently "fixed." Thus, sensory maps of the body that are located in *adults'* brains can be altered by experience. In a review of this astonishing work, J. H. Kaas wrote "recent experiments on the motor, visual, and auditory systems suggest that the capacity to reorganize characterizes all central representations and may be a general feature of brain tissue." He concluded, "sensory maps in a range of mammalian species have been shown to change. The clear implication is that adult plasticity is a feature of all mammalian brains."

But how is such plasticity possible? Kaas notes that some of the changes observed in primate brains after the loss of a digit "are so rapid, within hours, that . . . modification of existing synapses rather than growth of new connections is the likely explanation." He goes on to point out that "other modifications appear to take place over longer periods of weeks and perhaps months," suggesting that "sprouting of central axons may be an additional factor" needed to explain these changes.

More recent research indicates that less traumatic experiences, too, can affect the adult nervous system, at least in rats. Adult male rats, you will not be surprised to learn, will copulate in short order with receptive adult female rats, and not with similar but unreceptive female rats. Thus, males are significantly more sexually active when they are caged with estrogen-treated females who are, as a result of the treatment, constantly receptive; in contrast, males caged with continually unreceptive females rarely have sex with their cage-mates. Examination of the spinal cords of male rats caged in these varying circumstances has shown that their sexual experiences *as adults* significantly alter the size of their neurons. Thus, "differences in sexual behaviour [can] cause . . . differences in brain structure." Clearly, at least some of the things we experience as adults are able to leave their marks on our nervous systems, much as our experiences as children contribute to the later structures of our brains.

In fact, recent studies have found that the number of *new* cells generated in *adult* rats' brains is directly affected by experiential factors as well.

Specifically, new neurons created in a certain area of adult rats' brains have been found to be nearly twice as likely to survive if the rats live in complex environments than if they live in impoverished cages. Additional studies have demonstrated that simply giving rats the opportunity to run voluntarily on a running wheel is enough to support the survival of these newly created neurons. Other researchers studying this same brain area have found that "the number of adult-generated neurons doubles . . . in response to training on . . . [specific types of] learning tasks." Taken together, these results demonstrate that the "proliferation and survival of newly formed neurons can be affected by experience." In contrast to the now-discredited view that new neurons are never formed in adult brains, this research shows that new neurons *can* be formed in certain brain areas in adults (even human adults), and that both the number that are formed and the number that survive are affected by experience.

The emerging portrait painted by recent neuroscience research suggests that both mature and immature brains are structured, in a decidedly dynamic way, by the mutual activity of genetic and nongenetic factors. (Although I did not discuss in this last section the genetic factors that contribute to brain plasticity and structure, you can be sure that such factors are indispensable; new brain cells cannot be produced *or* survive without the integral participation of the genes.) The sensitivity of the nervous system to experience renders prespecification of brain structure and function impossible by any means, including via a genetic "blueprint." Instead, *at each moment of your life* some of the basic structural and functional characteristics of your brain reflect the contributions of your experiences, even if these events were experienced quite recently. Clearly, the particulars of these characteristics cannot be determined in advance of experience; they cannot be prespecified by the genes with which you were born.

10

TORNADOES, STARS, AND HUMAN BEINGS
The Developmental Systems Perspective

It is bad when one thing becomes two.

—From *The Hagakure: A Code to the Way of the Samurai*

D r. Ashborn Følling, a physician practicing in Norway in 1934, walked into an examining room to meet his next patients. Inside, he encountered a pair of blond-haired, blue-eyed children with severe mental retardation of unknown origin. The children's parents likely informed Dr. Følling that in addition to having cognitive deficits that seemed to have been present from birth, the children were prone to tremors and, on particularly bad days, to seizures as well. In the absence of any obvious course of diagnostic action, Dr. Følling did what most physicians would have done: he did a physical examination and ordered a battery of tests.

When Følling was finally able to examine the results of the tests he had ordered, he noticed something rather peculiar: both children's urine contained abnormally high levels of a chemical called phenylpyruvate (pronounced fe-nil-py-ROO-vate). No doubt other doctors had seen patients with this strange aberration, but perhaps they had dismissed it as a measurement error. Or perhaps Følling was helped to his observation by the fact that here in his hands were *two* tests reflecting the same abnormality in the urine of two individuals who appeared to be suffering from the same disorder. Regardless, no one before Følling had seen in this urine abnormality a potential key to understanding mysterious cases of mental retardation. But Følling took the obvious next step: he tested a number of other intellectually delayed children in Norway and, before long, he had discovered that many of them, too, had abnormally high blood levels of either phenylpyruvate or of a related protein, phenylalanine (pronounced fe-nil-AL-a-neen).

Building on this initial clue provided by Følling's careful observations, subsequent generations of researchers have made great headway in understanding the origins of this disorder, now known as PKU (short for *phenylketonuria*). It turns out that phenylalanine is an amino acid that we all consume in our normal diets; there are large quantities of it in milk, eggs, all types of meat, and bread. (It is also used to make aspartame—the generic name for NutraSweet—which artificially sweetens many diet sodas; this is why if you read the print on your diet cola can at lunch today, you'll see a warning that says "Phenylketonurics: contains phenylalanine"). The question, then, is why do some children with cognitive deficits have too much of this amino acid in their urine?

As it happens, normal human bodies produce a protein that chemically breaks down phenylalanine. Children with PKU do not produce this protein, so they fail to break down the phenylalanine they consume; consequently, abnormally high amounts of phenylalanine wind up collecting in their blood streams and urine. High blood levels of phenylalanine are associated with the development of mental retardation, seizures, tremors, and behavioral disorders. In addition, because one of the by-products of phenylalanine breakdown is the amino acid tyrosine, and because tyrosine breaks down to yield melanin (the pigment responsible for hair and eye color), PKU sufferers typically have abnormally *low* quantities of melanin in their bodies, and so blond hair and blue eyes. (This phenomenon, of course, reinforces the contention that coloration depends on more than simple genetic determination.*) But *why* do these children lack the protein that breaks down phenylalanine?

As you know, DNA can—and does—do only one thing: it provides information about the order of amino acids in a chain. But since proteins are amino acid chains, a particular cistron could conceivably carry information about the order of amino acids in the protein that breaks down phenylalanine. In fact, normal people *do* have a "gene" that codes for this protein, and children with PKU lack this gene. As a result, textbooks commonly describe PKU as being caused by a defective gene that leads to a failure to produce the protein needed to break down phenylalanine. Thus, PKU is understood by most medical doctors to be a classic example of a "genetic" disorder.

Were we to unquestioningly accept the conclusion that the cognitive deficits associated with PKU are genetic, we would have little hope of treating them other than attempting, somehow, to alter the DNA in every cell of PKU sufferers' bodies.* But our understanding that genes cannot single-handedly cause traits encourages us to consider how the genes of people with PKU *interact with their environments* to cause the *development* of these mental abnormalities. After all, genes only provide information about amino acid orders, and an abnormally functioning brain must be *built* during development via gene-environment co-actions. Understanding the situation in this way can open our eyes to developmental interventions that might be cheaper or more efficient than treatments relying on actual genetic manipulations. And, in fact, in this case, a cheap, effective, nongenetic treatment for this disorder was identified in 1954 by H. Bickel, J. Gerrard, and E. M. Hickmans. These researchers reasoned that since the symptoms of PKU result from an inability to break down a substance consumed from the *environment*, all we need to do to alleviate those symptoms is restrict the dietary intake of this substance in those people with this inability.

Given the severity of the mental retardation that develops in untreated cases of PKU, this is a serious disorder indeed. Moreover, PKU is the most common biochemical cause of mental retardation; as many as 2 percent of people with European ancestry are believed to carry the PKU gene, and as many as one out of every 10,000 babies born annually have the disorder. Amazingly though, simply limiting an infant's dietary intake of phenylalanine can *prevent* the cognitive deficits that characterize untreated PKU. In fact, if newborns at risk for PKU could be unerringly identified as such and subsequently fed an appropriately low phenylalanine diet, the severe cognitive deficits that normally characterize PKU *would not develop*. As a result, hospitals in industrialized nations now routinely test newborns' urine for abnormal levels of phenylalanine and subsequently restrict the phenylalanine intake of those who are at risk for PKU. In many cases, this treatment is successful.*

Even though PKU is widely considered to be a genetic disorder, I would argue that it in fact *develops* as a result of specific gene-environment interactions (of course, this is my argument whenever I am asked if a disorder is "genetic"). At first glance, it certainly *appears* that a defective genome causes PKU. However, the faultiness of this conclusion becomes apparent once we acknowledge that the severe mental retardation characteristic of this disorder *does not develop in the absence of either the genetic defect or of a diet containing phenylalanine*. Instead, it is possible to have the defective genome but still score within the normal range on generalized tests of intelligence. Thus, full-blown PKU cannot be caused by genetic factors alone, because the mere presence of a defective genome does not lead to this disorder except under specific—albeit very common—environmental circumstances, namely those in which normal quantities of phenylalanine are consumed in the diet. Nonetheless, even though everyone familiar with

PKU now understands the importance of diet in controlling the appearance of its symptoms, PKU is still usually labeled a "genetic" disorder.

As you may have gathered, my argument rests on a belief that the absence of certain symptoms constitutes the absence of certain disorders. I believe this, apparently along with much of the medical community—after all, we do not consider a person with HIV (the virus that "causes" AIDS) to have AIDS if he does not have the symptoms of AIDS. It is possible, of course, that each of our bodies might *normally* contain several genes that—in the right environmental circumstances—would contribute to the development of some unknown disorders. But if these environmental circumstances were almost never encountered during normal human development, the consequences of having these genes would almost never be felt. Surely, in this case, it would be inappropriate to say that we each *actually have* disorders of which we just happen to be blissfully unaware. By the same token, it seems unfair to *define* PKU as "lacking the gene needed for phenylalanine breakdown." Thus, *only individuals with symptoms* should be said to suffer from PKU. But if some individuals with the genetic defect in question nonetheless remain asymptomatic as a consequence of their *specific dietary experiences*, then it is misleading to call PKU a "genetic" disorder and thereby imply that it is caused by genetic factors alone. In fact, the severe mental retardation associated with PKU appears only in those people with genetic defects *who also have certain experiences*.

One might still wish to protest that *normal* diets contain phenylalanine and that *normal* people have genes that prevent PKU, so if people with abnormal genes develop the disorder in a normal environment, then it must be a genetic defect that causes the disorder. This is a reasonable argument, but it unnecessarily restricts the scope of our understanding. If we know that manipulating natural events sometimes alleviates suffering, then manipulating the environments of infants born with genetic abnormalities is clearly in order. But calling PKU (or any disorder) a "genetic" disease risks blinding us to the possibilities of manipulating *non*genetic factors in ways that might lead to more desirable developmental outcomes.

Here is another way to think about it. Between 1687 and the dawn of the twentieth century, the movements of all objects were understood to obey Sir Isaac Newton's laws of motion. Nobody realized that Newton's laws are not *generally* valid until Einstein showed that while they help us understand everyday phenomena, they do *not* help us understand the behavior of objects moving near the speed of light. In contrast to Newton's laws, Einstein's theory of relativity accurately predicts the behavior of ordinary objects *and* the behavior of very fast-moving objects. Similarly, we *can* *think* of PKU as being caused by genetic defects that cause PKU, but this "understanding" is useful only *in specific (normal) circumstances*. Finding possible nongenetic solutions to developmental problems requires having a *general* understanding of how genes and environments interact to produce traits, an understanding that is valid in abnormal as well as normal environments. After all, some abnormal environments can be therapeutic, as phenylalanine-poor environments are for people unable to break down phenylalanine! Finding a solution to a problem is most likely if one carries around in one's head as complete an understanding as possible of the factors that contribute to the problem. This is why *heritability* information—which "accounts" for trait variation *in specific situations*—is never useful in the way that *developmental* information is; only the latter produces the kind of complete understanding that has *general* utility.

DEVELOPMENTAL SYSTEMS

Recognizing the utility of an approach that seeks to understand trait development *in general* (as opposed to in "normal" environments only), many developmental scientists have begun to view development as being driven by the co-action of components that constitute a single integrated *system*. Perhaps a preliminary appreciation for this idea can be elicited by a metaphor. Imagine a piece of spontaneously produced instrumental music—a jam—in which none of the players is carrying the melody, in which each player is playing something different from the others but fascinatingly beautiful in its own right, and in which each of three musical lines are all—blissfully—rhythmically and melodically compatible. Presumably, spontaneous harmony like this could arise only if the players were listening closely to one another while they were playing, altering their notes at the spur of the moment, so as always to complement one another. The contributions of the three players are equally important, and, in fact, if one of the players ceased playing, the piece might even become unrecognizably different. Many developmentalists now think that the contributions of genetic and nongenetic factors to our traits are like the contributions of these artists to their musical composition. The analogy has its shortcomings, of course (like all analogies): if one musician quits, the others can keep playing, whereas the complete *absence* of either chromosomes or an environment leads to a complete absence of a developmental outcome. Nevertheless, this metaphor works in at least two important ways. First, each of the players is necessary (but not sufficient) to produce the specific piece of music. Second, as long as the players are dynamically responsive to one another—building the music together as they play in real time—none of them can be thought of as the leader or as more important than another player; instead, they are all merely *collaborators*. Notwithstanding popular confusion about such things, the facts of molecular biology and embryology are consistent with the idea that traits develop like our piece of music: genetic and nongenetic factors constitute a single dynamic system within which these integral components interact in real time to construct traits. Let's examine this idea in a bit more detail.

According to the developmental systems perspective, animals are made of components that exist on several levels of a hierarchy. These components include the genes, the genes' environment (the material within a cell that surrounds the genes), the cell's environment (other cells that collectively constitute an organ), the organ's environment (a collection of other organs which together, constitute an organism), and the organism's macro-environment. Thus, our genes are embedded in a hierarchical series of environments, some of which are in closer proximity to the genes than are others, but *all* of which constitute aspects of the genes' environment. Moreover, developmental systems theorists maintain that *each of the system's components can interact with, and affect, each of the other components.*

As these theorists see it, in addition to components on a particular level of the hierarchy affecting one another—as when genes activate other genes, or when cells induce the differentiation of other cells—components on different levels of the hierarchy can affect one another as well. Between-level interactions are seen, for example, when genes contribute to the production of proteins that confer the characteristic structure of differentiated cells or when the structures of cells affect the structures of organs. According to the developmental systems perspective, *what occurs at one level of a developing system is never isolated from what occurs at another level.*

One of the distinguishing features of this perspective is its insistence that components of a system influence one another *bidirectionally*. This means that components on "lower" levels (for example, genes) can affect components on "higher" levels (for example, cells) and that components on "higher" levels can affect components on "lower" levels. For example, we have seen how alternative RNA splicing allows different genes to be "activated" as a function of the type of cell doing the splicing and also how hormones circulating in the blood can affect the activity of the genes. Such effects of "higher" level components on "lower" level components are common, but traditional discussions of the roles of nature and nurture in trait development never acknowledge how regularly "higher" level activity dramatically affects "lower" level activity.

Another feature of the developmental systems perspective—a feature as important as bidirectionality—is its tenet that the macroenvironment is just another component of the developing system. That is, like every other system component, the macroenvironment is understood to be a component that is bidirectionally and integrally involved in driving development and constructing the system's traits. This idea is so unconventional that had I mentioned it earlier, it might have been met with disbelief. But now, I hope several examples will come to mind of development-altering interactions between environmental and biological components of developing systems. For example, recall that newborn rat pups have glands that secrete a chemical that elicits maternal licking; processes in a pup's organs clearly affect events occurring in the pup's environment—in this case, the pup's mother—and events at this "higher" level certainly influence the pup's subsequent development. Conversely, events at "higher" levels can affect events at "lower" levels; recall the activation of human immediate-early genes in response to stressful academic examinations.

Recent studies—for example, on the development of walking in human babies—have demonstrated the value of construing each human being *and* his or her environment as a *single* integrated dynamic system. When we view people and their environments in this way, we effectively do away completely with the age-old person-environment dichotomy that most of us intuitively maintain. Because such a nondichotomous approach treats organisms as inextricably embedded in their environments, you and your environment—together—are understood to make up a *unitary* system that must be studied as such. This understanding brings with it some profound implications, of course, not the least of which is the insight that altering *any* of the components of the system can influence development. It is this insight that gives this approach its generality and encourages us to seek environmental solutions to so-called genetic disorders like PKU.

The idea that biological factors do not operate independently of environmental factors—that these factors are all necessary interacting collaborators in the processes that produce traits—is quite different from traditional conceptions that sharply distinguish between the roles genes and environments play in development. Nonetheless, this idea has begun to appear in the writings of several theorists. For example, it was clearly stated in the introduction to an article that the philosophers P. E. Griffiths and R. D. Gray wrote on this topic:

> The genes are just one resource that is available to the developmental process. There is a fundamental symmetry between the role of the genes and that of the maternal cytoplasm, or of childhood exposure to language. . . . There is nothing that divides the resources into two fundamental kinds. The role of the genes is no more unique than the role of many other factors.

Later, these theorists write:

It makes no more and no less sense to say that the other resources "read off" what is "written" in the genes than that the genes read off what is written in the other resources. The reading of the genes is a metaphor which has been of some historical utility, but which now retards the study of development, and . . . evolution.

According to this perspective, then, genetic and nongenetic factors have comparable roles in the development of our traits.

Snowflakes, Babies, and Ecosystems: Self-Organization

Once the components of a developing system are viewed in this way, genetic and environmental factors both lose their "special" status as drivers of development. But if neither of these factors alone can drive development, what can? Developmental systems theorists view biological traits as *emerging* from the interactions of two or more collaborating, integral components of a system. Gottlieb, Wahlsten, and Lickliter put it like this: "New structural and functional properties and competencies [emerge] as a consequence of . . . co-actions among the system's parts." An essential aspect of this perspective is that no single component of the system is believed to contain detailed *instructions* for building a body's (or brain's) characteristic traits; instead, traits are understood to develop as they do because of the collective activity of *all* of the system's components. But how can this be? How can interactions between a system's components give rise to novel forms—as when the rudiments of a brain begin to appear from a previously undifferentiated mass of cells or when language begins to emerge from a babbling infant—how can these new forms arise if no one of the components of the system has instructions about how to construct them?

Potential answers to this question have surfaced in the wake of recent research on *non*biological, but still complex, dynamic systems.* Specifically, work in physics and mathematics has shown that under certain circumstances, physical systems can *spontaneously organize themselves*. What this means is that they can settle into states with characteristic properties—traits—that emerge over time (that is, they develop). The behavior of these systems is nonrandom, and the systems themselves have characteristic shapes and internal structures. Nonetheless, no instructions for these characteristics preexist in the interacting components that constitute the system and that, collectively, give rise to the system's properties.*

In the spring of 1999, Americans watched in horror as a series of devastating tornadoes touched down in the heartland, bringing death and destruction in their wake. When I saw news video of these twisters, I was struck by what an excellent—if horrific—example of self-organizing systems they were. Clearly, "instructions" for constructing a tornado do not exist anywhere in nature (at least, not in the way that people have traditionally thought of "instructions" for building a person somehow existing in DNA). Nonetheless, here was an organized phenomenon that *emerged* out of the interactions of the many, many components that constitute complex weather systems (including the molecules that make up the air out of which tornadoes are constructed, temperature differences that exist at different altitudes, prevailing winds, and so on). Under the right circumstances, these uninstructed, inanimate components can organize themselves into devastating weather systems with specific nonrandom properties. For example, tornadoes have characteristic traits, including their tendency to rotate in a counterclockwise direction in the northern hemisphere. Like- wise, some remarkably organized "traits" distinguish various types of thun- derstorms, including their characteristic and well-defined life cycles and life spans, the tendency of certain types of storms to be nocturnal, and in the case of so-called supercells that are known to be prolific breeders of tornadoes, their tendency to move to the right of prevailing winds. Esther Thelen and Linda B. Smith write:

> In certain meteorological contexts, clouds form into thunderheads that
> have a particular shape, internal complexity, and behavior. There is a clear
> order and directionality to the way thunderheads emerge over time . . .
> just as there is in [individual biological] development. But there is no
> design written anywhere in a cloud. . . . There is no set of instructions
> that causes a cloud . . . to change form in a particular way. There are
> only a number of complex physical . . . systems interacting over time,
> such that the precise nature of their interactions leads inevitably to a
> thunderhead. . . . We suggest that action and cognition are also emergent
> and not designed.

Weather systems are not extraordinary instances of such self-organization; a list of self-organizing systems would include stars, snowflakes, and ecosys- tems, to name but a few common examples.*

These examples indicate that interactions between components of com- plex systems sometimes give rise to systems with characteristic behaviors, shapes, and internal structures, each of which develops over time. As a result, it is not far-fetched to believe that the characteristics of biological systems—our traits—emerge in the same way, that is, from interactions among the system's components. When Nobel laureate Edelman writes, "The brain is . . . a self-organizing system," he is acknowledging the value of thinking about biological systems using these new ideas that have rev- olutionized how we think about dynamic *physical* systems. Given what we now understand about the emergence of complex structure from inani- mate, uninstructed, and interacting components of dynamic systems, it is reasonable to believe that biological development, too, can arise from interactions, in this case among genes, their immediate environments, cells, organs, organisms, and the broader macroenvironment. Biological systems—people—then can be understood as natural phenomena that, like thunderstorms and stars, emerge and develop over time, and in specific circumstances.

"Traits" can develop spontaneously in physical systems even if instructions for the traits are not located in any one of the system's components; similarly, "instructions" for biological traits need not reside in any one of a biological system's components. Instead, the "information" required for the construction of biological traits can be *distributed* across all of the components of the developing system. Thus, in this case, each component would be necessary but not sufficient for the appearance of the trait. Under these circumstances, one component of the system cannot possibly be more important than another in determining a trait's appearance. Rather, traits develop as they do because of the co-actions of *all* of the components that constitute the developing system, including the environment in which development is proceeding. In Gottlieb's eloquent words, "the cause of development—what makes development happen—is the relationship of the . . . components, not the components themselves. Genes in themselves cannot cause development any more than stimulation in itself can cause development."

This analysis suggests that our inability to measure the relative importance of genetic factors to the development of particular traits is *not* due to a lack of progress in molecular or developmental biology (or any other science). Instead, technological advances will not resolve this question, ever; causal contributions to our traits cannot be apportioned to the components of the system from which we emerge, because *the very nature of the developmental processes that build our traits renders apportioning causality theoretically impossible*. As Gottlieb, Wahlsten, and Lickliter state in their enlightening chapter in the *Handbook of Child Psychology*, "the question of [which component] is more important for . . . development is nonsensical because both are absolutely essential."

From this vantage point, then, the type of questions we should ask about the source of our characteristics must change from "Is this trait determined more by genetic or environmental factors?" to "How do the components of a system co-act *in development* to produce a trait?" All of our physical and psychological characteristics reflect the fact that they *developed* at *some* point in our lifetimes; after all, traits do not appear out of thin air! Our bodies—and, therefore, all the things our bodies do, from breathing to playing a Bach fugue on the piano—are *built* over time by the mutual, collective activity of our genes, their microenvironments, and our macroenvironments. Hence, the only way to "understand the origin of any [trait is] . . . to study its development in the individual."

The developmental systems perspective is not an entirely new way to think about development. In fact, such an approach has necessarily been with us since Driesch separated the cells of a two-celled sea urchin embryo and watched in disbelief as each cell developed into a complete sea urchin. How else can we understand the bombshell-like finding that if two cells of an embryo are left attached, the pair gives rise to a single complete animal, but if the two cells are separated, *each one* gives rise to a complete animal? The only conclusion that can be drawn from this finding is that the outcome of development cannot possibly be predetermined by factors solely within the zygote (or within its environment, for that matter). This result fairly screams that context is crucially important in development; developmental outcomes do not result from context-insensitive "translations" of instructions contained in a zygote. Biology simply does not work that way.

LODGEPOLE PINECONES AND STRANGLER FIG SEEDS: LAYING IN WAIT FOR INFORMATION

On further reflection, it is hard to imagine how development could arise in the *absence* of environmental events; it is not as if disembodied genes *ever* spontaneously become active without some environmental event starting the cascade in the first place. One of nature's phenomena that I think drives home this point is called "dormancy." Organisms (or reproductive bodies within organisms) are said to be dormant when they are in a state of developmental inactivity that ends only under certain environmental conditions; such a state occurs commonly in the life cycles of various plant species (but is also seen in animals). Many seeds—peas are a good example—can germinate after they are harvested, but they do not unless they are kept moist. The more exotic example of lodgepole pinecones is particularly fascinating. T. Hackett notes:

> It is not merely that the lodgepole-pine forests in Yellowstone have
> tolerated fire: they are a product of it. The forests have evolved through
> a pattern of mass fires occurring at intervals of between two and four
> hundred years. The saving grace for many lodgepole pines is their
> serotinous cones, which normally are closed tightly and are as hard as
> rocks, but open up when they are subjected to intense heat, such as fire,
> and sow seeds in the ashy, fertilized soil after a burn.

Hackett goes on to observe that "in a typical burned acre, where perhaps three hundred and fifty old lodgepoles once stood, perhaps a hundred thousand pine seeds were scattered. . . . Within five years some six thousand saplings will rise from the forest floor." After the 1988 fires in Yellowstone National Park, pinecones that had fallen to the forest floor and had laid dormant for many years suddenly "came to life," producing new saplings; a similar process is taking place now in Los Alamos, New Mexico, which is recovering from its own recent firestorms. It is clear in this case, of course, that the environment is a source of "information" for the pinecones: had the pinecones germinated, somehow, before a fire, they would not have survived in the shade of the parent pines, competing with these older pines for soil nutrients and other resources. In fact, the fire served as an effective information-laden signal that the time was ripe for successful development.*

I learned of another interesting example of dormancy on a recent sabbatical trip to Costa Rica. There, a particular species of parasitic fig tree—locally called a strangler fig—sprouts roots in the boughs of another (host) tree. These roots grow down around the outside of the host tree's trunk, ultimately growing right into the earth below. In time, the fig grows all around the host tree, finally strangling it. By the time the host tree is dead, the fig has grown strong, and when the host tree finally decays away, a perfectly hollow fig tree is left behind (perfect for tourists who wish to climb up inside the trunk into the rain forest canopy above). But how do the fig's seeds get into the canopy in the first place, where their development *must* begin if their dastardly plan is to work?

It turns out that strangler fig seeds remain dormant until such time as they have passed through the digestive tract of a primate, and not before. Thus, their development begins after the fruit that contains them is eaten by a monkey, who, after pooping in the treetops, effectively leaves the seeds in the required location. In this case, the environmental stimulus that sets the developmental cascade in motion is something within the gut of the monkeys that eat these figs. Here again, the environment clearly plays an "informational" role in the fig tree's development; the presence of the environmental stimulus effectively "informs" the seed that in all likelihood, it will soon find itself in a treetop, the only location from where its development can salubriously proceed. Nonetheless, "instructions" specifying the characteristics of the mature tree cannot be found exclusively either in the environmental stimulus or in the chromosomes contained in the fig's seeds.

RESISTANCE AND SUPPORT: THE STATUS
OF THE DEVELOPMENTAL SYSTEMS PERSPECTIVE

The notion that development depends on the contexts in which development occurs is an old idea. But even though understanding development ultimately *requires* adopting a systems perspective, much of the public continues to believe that some psychological and many biological traits are "largely genetic." Why might we retain this belief in the face of evidence suggesting that it is false? I can offer three possible hypotheses, reasons that are not mutually exclusive (or exhaustive). First, the contemporary public's willingness to accept the idea that traits are differentially influenced by genetic factors might reflect the fact that the truth is significantly more complicated than the fiction, making it difficult to comprehend. Second, it may be that the facts needed to understand trait development have not been adequately conveyed to the public by those responsible for the public's education; this would include college professors, secondary-school teachers, and those who produce our newspapers and television shows (among others). Third, it might be that our perceptions obfuscate the facts. Perhaps casual observation of development sustains beliefs that are incompatible with scientific observations, much as casual observation of celestial body movement led the sixteenth-century public to believe that the earth was the center of the universe, even as the scientific observations of the day began to provide conclusive evidence to the contrary. Regardless, it is clear that dislodging the idea that some traits are "primarily genetic" has not been easy.

The good news is that the developmental systems perspective is now "being ever more widely used in developmental psychology" and, in fact, is being increasingly integrated into mainstream thought underlying *all* of the life sciences. This observation is supported by the fact that the Royal Swedish Academy of Sciences, which chooses most of the winners of Nobel Prizes each year, sponsored a 1994 symposium in Stockholm that was organized around this theoretical framework. Perhaps a new day is dawning.

Ultimately, this shift in perspective necessarily bears on how we think about evolution. The "popular" understanding of evolution holds that traits that are both inheritable and that allow individuals to produce greater numbers of fertile offspring will be more common among surviving offspring than traits that do not advance this cause (or that are not inheritable). Because chromosomes are commonly thought to be the only biologically significant factors that offspring inherit from their parents, genes are often assumed to single-handedly cause the evolutionarily adaptive traits that our ancestors have bequeathed to us. But if the ideas I have presented so far are correct, and traits—whether adaptive or not—are caused by genetic factors that *depend for their functioning* on nongenetic factors, our understanding of evolutionary processes will have to be revised.

Part IV

DEVELOPMENT AND EVOLUTION

We're all of us guinea pigs in the laboratory of god. Humanity
is just a work in progress.

—*From* Camino Real *by Tennessee Williams*

11

THE MODERN SYNTHESIS
What Mother Nature Selects

On the day my friend's daughter Callie was born, I heard another friend marveling: "She has her father's lips!" Sure enough, there was something about Callie's lips that was definitely reminiscent of her father's. But no doubt Callie didn't have her father's lips; Callie had her own lips! How can Callie have been born with lips that bore such a close resemblance to her father's? In what sense did she *inherit* her father's lips?

The issues underlying these questions go to the heart of the relationship between development (which happens to an individual over a lifetime) and evolution (which happens to a species over generations). On one hand, Callie was born with the lips she was born with because she *developed* them sometime between conception and birth. On the other hand, their close resemblance to her dad's lips suggests that Callie's lips were somehow *inherited* from her father, via a process that is more evolutionary than developmental. Is there a relationship between the processes of development and evolution, and, if so, what is its nature? And what, if anything, does adopting a developmental systems perspective do to our ideas about how these phenomena are related?

Long before Charles Darwin offered the first credible nonsupernatural explanation for the existence of humankind, some people had thought about how individual development is related to the succession of living things that runs from the simplest plants to the most complex animals. But when Darwin showed that species *evolve* over generations, parallels between individual development and evolution could no longer be ignored. By 1874, Ernst Haeckel had successfully popularized the idea that an individual's development is a rapid repetition of the major changes in bodily form undergone by its ancestors as they evolved across generations. Among the more compelling bits of evidence marshaled in support of this idea were the facts that human embryos—like ancestral fish—possess gill slits, and that as a human heart develops in utero, it first resembles the heart of an insect, then the heart of a crustacean, then the heart of a fish, then the heart of a reptile, and only finally the heart of a human. These sorts of observations led most late-nineteenth-century biologists to believe that the development of individual organisms can be understood to be an actual recapitulation of the evolution of their species.

Biologists have since rejected the idea that development recapitulates evolution; no one believes this any more.* Even so, biologists still believe that there is a very close (and important) relationship between the two phenomena. Thus, while it is certainly possible to consider questions about development independently of questions about evolution, I have decided to consider evolution in this book nonetheless.

The need to consider evolution in a book about development arises from the fact that various forms of Darwinism now inform the thinking of the scientifically literate public. Specifically, we often imagine that natural selection acting in previous generations delivers to each new baby a set of genes that then determine the baby's traits. Embedded in this vision is the idea that certain traits—such as 10 toes in human babies or so-called behavioral instincts—are genetically determined. Developmental systems theory, though, tells us that this is not possible, because traits are caused by the co-action of genetic and nongenetic factors, *both* of which carry "information" required for trait development. So, how can this perspective be reconciled with that of evolutionary biology? Doesn't evolutionary biology insist that natural selection gives us adaptive traits by ensuring that "information" sufficient to produce these traits is encoded in our genes? I hope that examining the relationship between development and evolution will help readers refrain from reverting to genetic determinism when faced with the sorts of apparent challenges to the developmental systems perspective that arise from fields such as evolutionary biology, sociobiology, and evolutionary psychology.

THE PARTY LINE AND ITS SHORTCOMINGS

Today's scientifically informed public has an accurate understanding of natural selection. Specifically, the idea as Darwin originally conceived of it is that variability in a population's inheritable traits, coupled with limited resources for which individuals must compete to survive and reproduce (for example, food, mates, and shelter), ensures that inheritable, adaptive traits—those that increase the likelihood that an individual will produce fertile offspring—will be more likely than other traits to characterize the descendants of the population. Unfortunately, this understanding, while correct, masks some underlying ambiguities that become apparent only when additional thought is directed toward the problem. Among these ambiguities are those raised by the following questions: What, exactly, is it that nature "selects"? What does it mean to say that a trait is "inheritable"? How do the variations among which nature selects appear in the first place? And, finally, what is the nature of the relationship between individual development and evolution?

It turns out that while much of the educated public believes that natural selection operates on genes per se, this idea cannot be reconciled with what is known about biological development. The following fanciful example will help to illustrate this point. Imagine that a particular fish in a school has a genetic mutation that contributes to the development of "deformed" fins, fins that allow her to crawl up onto the sand once she finds herself in sufficiently shallow water (humor me by letting her have air-breathing lungs as well). The next time a shark chases the school toward the beach, our fish will discover that she has an advantage over her normal-finned peers: as they die by the dozens in the shallows, she will just "walk" away on the sand, thereby surviving another day and increasing her chances of living long enough to reproduce. Of course, when our heroine finally does reproduce, her offspring will have some of her genes. In contrast, the genes of the other fish in the school—genes that have unexpectedly become shark food—are no longer contenders on the evolutionary battlefield. This is the picture of evolution widely held by today's public: by destroying the genes of the normal-finned fish but not the genes of the walking fish, nature has effectively "selected" the walking fish's genes for continued survival. The typical ending to this story, then, reads: "As a result of being the only fish of her species to survive repeated shark attacks, future generations of our heroine's species will possess her genes and, therefore, be able to walk." It is in this sense that natural selection is thought to operate on genes.

The problem with this fish story is its underlying assumption, namely that the genes preserved are genes "for" walking. Is such an assumption warranted? To address this question, we need to first consider what, exactly, we mean when we say there are genes "for" specific traits. In common English, when we say that the pedals on a bicycle are "for" making it move, we do *not* mean that pedal pushing, *in and of itself*, always causes the bicycle to move; we understand that bicycle movement also depends on the bicycle having an intact chain and appropriately positioned round wheels, among other things. Similarly, when biologists say that there are genes "for" specific traits, they do *not* mean that these genes, *in and of themselves*, cause these traits to develop. Biologists know that genes cannot cause traits to develop independently of other factors.

Unfortunately, the claim that there are genes "for" specific traits ultimately misleads the rest of us. The confusion arises from the fact that when we hear there is a gene "for" a specific trait, we usually (mis)take such a statement to mean that the gene *alone* causes the trait, independently of the contexts in which development occurs. If a typical mother is told that her baby has "the gene for Wilson disease," she will undoubtedly conclude that her child *will* eventually develop the symptoms of this disease. Fortunately, such a conclusion would be a mistake; as is the case with PKU, there are easily produced developmental circumstances in which individuals with "the gene for Wilson disease" can wind up indistinguishable from normal people.*

Similarly, the epigenetic nature of trait development makes it likely that the same genes that helped our mutant fish develop deformed fins could, in different contexts, contribute to the development of normal fins. Ultimately, genes *can* be "for" traits in the same way that bicycle pedals can be "for" locomotion. But think with care about what this means: if the bicycle in question is on Gilligan's Island, those same pedals might be "for" generating electricity to run the radio, not "for" locomotion at all. By the same token, genes "for" deformed fins in some developmental circumstances could be genes "for" normal fins in other circumstances. Thus, to the extent that calling a gene "the gene for deformed fins" implies that these genes cause deformed fins *in all developmental circumstances*, calling them genes "for" deformed fins is misleading.

At this point, we begin to see the problem with the presumed ending of our fanciful fish story. Implicit in this ending was the idea that the genes preserved when our mutant heroine outmaneuvered the attacking shark were genes that would invariably give her offspring her "deformed"—but adaptive—fins. But, in fact, there are no such things as genes that produce specific traits in *all* developmental contexts, so our heroine's genes alone cannot provide her offspring with her novel adaptive trait.

The most important implication of this conclusion is that natural selection probably does not operate on genes *alone*. Why? Because natural selection would not be very effective if it worked by selecting genes that *might—or depending on the disposition of a variety of other factors, might not—* contribute to the development of adaptive traits. Thus, since genes *never* absolutely specify traits, natural selection is unlikely to operate on genes alone. In the end, the idea that natural selection operates on genes alone fails, because it omits from the picture the *development* that characterizes the lives of all organisms.

But what else can nature select? As it happens, a ready answer to this question lies in orthodox Darwinism. After all, remember that when Darwin originally formulated his theory, the modern conception of the gene was not yet even a twinkle in anyone's eye. Instead, Darwin understood that the *mechanism* by which selected traits develop does not bear on his theory at all. All that matters is that *some* mechanism of inheritance ensures that adaptive traits are transmitted faithfully across generations. Given the existence of such a mechanism, if a walking fish is less likely than a nonwalking fish to become a predator's lunch, then the descendants of that fish effectively inherit its ability to walk, not genes "for" walking, per se.* As Darwin reckoned it, *the actual process by which walking fish develop the ability to walk is of no concern to nature as it "selects" those fish that will populate the next generation; all that matters is that traits that contribute to survival and reproduction are invariably transmitted to the next generation, somehow.* In this way, natural selection can be seen to operate not on genes, but on traits themselves. Put concisely by the famous biologist Ernst Mayr, "natural selection favors (or discriminates against) phenotypes,* not genes." The idea that *genes* provide the needed mechanism of inheritance was added years after Darwin's death, in a theoretical formulation (forged, in part, by Mayr himself) known as "the modern synthesis." But this latter idea, discussed below, was never central to Darwin's insights about natural selection.

"PANGENESIS": DARWIN AND THE INHERITANCE OF ACQUIRED CHARACTERISTICS

There is a tale underlying the widespread popularity of the idea that natural selection operates on genes alone; it can be told beginning with Darwin's belief that the characteristics organisms acquire over their lifetimes—as a result of their experiences—can be passed down to their offspring. Most biologists have since rejected the idea that acquired traits can be inherited,* but this idea was not at all controversial in Darwin's time. In fact, for literally thousands of years before Darwin, virtually everyone *presumed* that traits acquired in one's lifetime could be passed on to one's children. Thus, it is not surprising that Darwin thought the inheritance of acquired characteristics was perfectly compatible with his posited process of natural selection.* On page 4 of *On the Origin of Species*, he wrote:

> Changed habits produce an inherited effect, as in the period of the
> flowering of plants when transported from one climate to another. With
> animals the increased use or disuse of parts has had a more marked
> influence; thus I find in the domestic duck that the bones of the wing weigh
> less . . . than do the same bones in the wild duck; and this change may be
> safely attributed to the domestic duck flying much less . . . than its wild
> parents. The great and inherited development of the udders in cows and
> goats in countries where they are habitually milked, in comparison with these
> organs in other countries, is probably another instance of the effects of use.

But what sort of mechanism could support the inheritance of acquired traits? Darwin favored a version of a theory that had been around in one form or another since the time of Hippocrates. The basic idea was that sometime prior to reproduction,

> all parts of the body sent small representative particles to the reproductive organs where they formed "the germ," which gave rise to the next generation. . . . In the version elaborated by Darwin, the particles were christened "gemmules" and the whole hypothesis "pangenesis." According to Darwin . . . an environmentally modified [body] part, or a part that had become modified as a consequence of use and disuse, liberated modified gemmules into the circulation. The modified gemmules reached the germ cells [in the genitalia] and eventually participated in the formation of the corresponding modified [body] part in the offspring. In this way, acquired characters could be passed on to the next generation.

Thus, Darwin's belief in the inheritance of acquired characteristics via a pangenetic mechanism was a prominent part of his original theory of evolution.

WEISMANN'S BARRIER

Then, at the end of the nineteenth century, August Weismann—who was an ardent Darwinian but who was nonetheless skeptical that acquired traits could be inherited—questioned Darwin's decision to feature this belief in his theory. In an effort to evaluate this idea, Weismann conducted a series of experiments in which he cut off the tails of newborn mice, mated them with one another when they reached maturity, and then looked for evidence that their offspring had inherited the tailless trait "acquired" by their parents earlier in their lives. After mutilating 22 successive generations of mice and still finding that their offspring were born with healthy tails, Weismann put forth a doctrine he called "the continuity of the germ plasm." Specifically, this doctrine stated that the germ cells (that is, the sperm and egg cells) that contribute to an individual's offspring are protected from changes that occur in the somatic cells (the rest of the body's cells) during an individual's lifetime.* That is, modifications to body parts do not affect germ cells at all, rendering the inheritance of acquired characteristics through a pangenetic mechanism an impossibility.

By the beginning of the twentieth century, most biologists had begun to believe in the existence of Weismann's Barrier, a theoretical, impermeable boundary between the somatic cells and the germ cells. Weismann and his followers were not surprised, then, when W. E. Castle and J. C. Phillips first reported in 1909 that black guinea pig pups can be born to a white guinea pig mother if her ovaries have been surgically replaced with a black female's ovaries prior to conception. Given an impenetrable barrier of the sort proposed by Weismann, the fact that *all* of the somatic cells in a white mother's body are "white" is irrelevant: her new, surgically received germ cells are "black" and cannot be affected by somatic factors "beyond the germ line." Given her transplanted "black" ovaries, the white mother is, for all *reproductive* intents and purposes, black.

The repercussions of accepting the broad conclusion that acquired characteristics cannot be inherited are hard to overstate. In particular, embracing this conclusion had one *very* far-reaching consequence for twentieth-century biology: it effectively led to the conceptual separation of evolution and development.* Why is that? Because if events occurring over the course of our lives—as we *develop*—have no bearing on the traits of our offspring, developmental phenomena cannot possibly have a role in evolution by natural selection; this conclusion follows from the fact that natural selection works only on *inheritable* traits, and Weismann's data suggest that developmentally acquired traits cannot be inherited. A century later, the conceptual separation of evolution and development has led to the widespread belief that our traits are determined by *either* events experienced during our lives *or* by factors that affected our ancestors before we were even conceived. Accordingly, adaptive traits common to all members of a species are often thought to owe their existence to evolution and to be genetically determined. In contrast, our individual, unique traits are often thought to owe their existence to the idiosyncratic events that we experienced as we developed. Even though such an either-or scenario does not fit with the known facts of developmental biology, such is the legacy of accepting Weismann's conclusion that traits acquired as a result of experience cannot be inherited.

SHE HAS HER MOTHER'S EYES? JOHANNSEN DEFINES "HEREDITY"

Putting the genetic, developmental, and evolutionary pieces of this rather large puzzle back together again can be accomplished by revising how we think about what it is that is passed down to us from our parents. Our contemporary understanding of biological heredity can be traced largely to the efforts of early-twentieth-century biologists to distinguish between the *material* that is passed from parents to their offspring and the *traits* that were assumed to be *determined* by that material. A milestone in the conceptualization of heredity came in 1911 when Johannsen—two years after he coined the word "gene"—published a paper entitled "The Genotype Conception of Heredity." In this paper, he argued that biological heredity does *not* involve the transmission of traits, as had been assumed by everyone since antiquity (and by many nonbiologists today). Instead, all that can be transmitted—actually transferred—from parents to children are the parents' gametes—the sperm and egg. Given this conceptualization, Johannsen wrote, "Heredity may then be defined as *the presence of identical genes in ancestors and descendants.*"

According to this definition, a child's traits cannot be *inherited* from her parents. Instead, descendants can only inherit genes. A child's *traits*, Johannsen argued, *develop* much as her parents' traits developed during *their own* childhoods, that is, as a function of the "reactions" of the gametes to their "conditions." Thus, Johannsen would have maintained that if my friend's daughter Callie has her dad's lips, it is not because she inherited his lips, but because she *developed* lips like her dad's in much the same way as he originally developed them three decades earlier.

Johannsen's rather narrow concept of heredity—genes alone are inherited—was the one ultimately adopted by Americans working in the then-new field of genetics. Thereafter, "as genetics increased in importance and influence, so did this view of heredity." As a result, Johannsen's definition, like the idea that acquired characteristics cannot be inherited, had extraordinary repercussions on subsequent intellectual developments in biology. Although Johannsen knew quite well that traits are not caused by genes operating independently of developmental "conditions," his narrow conception of heredity inadvertently contributed to the idea that inheritable traits are determined strictly by genes. Together with Morgan's contemporaneous conclusion that chromosomes are responsible for the development of inherited traits and Weismann's insistence that acquired characteristics could *not* be inherited, Johannsen's notion of heredity laid the groundwork for the neo-Darwinian "modern synthesis" that was forged in the late 1930s and that is still very much with us today.

THE NEO-DARWINIAN MODERN SYNTHESIS

The modern synthesis—so named because it synthesized Darwin's theory of evolution by natural selection and the theory of the gene that emerged from the pioneering work of Mendel, Weismann, Johannsen, and Morgan—is the crowning achievement of neo-Darwinism. Briefly stated, the modern synthesis holds that evolution occurs when the frequencies of certain genes in a population of organisms change as a result of natural selection. This theory continues to offer the most agreed-upon answers to questions about the nature and origin of species; it is commonly accepted by biologists as *the* theory of evolution. And therein lies the reason for the widespread acceptance of two ideas: first, that natural selection operates on genes alone, and second, that naturally selected genes single-handedly cause the development of traits that have proven adaptive during the course of evolution.

The problem with the modern synthesis is that, at its core, it is strictly a nondevelopmental theory of genes, because it views genes alone as "the material basis of evolutionary change." What this means is that data from developmental biology are conspicuously absent from the modern synthesis, even though developmental phenomena have rightfully played a central role in shaping other biological theories since antiquity. The fact that the modern synthesis has little to say about development and differentiation should surprise no one: the modern synthesis *has* to be a nondevelopmental theory, since it was built on fundamental assumptions that are irreconcilably at odds with the fundamental realities of biological development. The problem is that while the modern synthesis allows that inheritable traits can be determined by genes alone, the data of developmental biology affirm that *all* traits emerge from epigenetic *co-actions* of genetic *and* nongenetic factors. Clearly, something has to give. The exclusion of developmental data from the modern synthesis continues to interfere with efforts to forge a comprehensive theory of evolution.

Were the architects of the modern synthesis ignorant of the importance of development? In most cases, not at all; in fact, most of them understood quite well the extraordinary thorniness of this problem. As a result, they simply chose to finesse the issue by defining evolution explicitly as a process affecting *populations*, not individuals; as we saw earlier, theories about the genetics of populations can ignore questions about development entirely.* This move, then, allowed the architects of the modern synthesis to forge a working theory of the genetics of evolution. But, by ignoring development in their explanation, their theory succeeded at the cost of a comprehensive account of the appearance of *any* of our traits. Later, as the twentieth century wore on and the ideas underlying the modern synthesis trickled down to the public, we—the public—wound up falsely believing that the theory of evolution can satisfactorily explain the appearance of individuals' traits. In place of an understanding that this theory can do nothing of the sort, we were left with the assertions that we inherit nothing from our parents but our genes, that natural selection operates on genes alone, and that genes can single-handedly determine the form of evolutionarily adaptive traits.

There is at least one serious problem with the idea that we inherit nothing from our parents but our genes. How can we reconcile Darwin's insight that evolution occurs only when *traits* are passed from generation to generation with Johannsen's conclusion that *traits per se* are not inherited but rather *develop* as genes "react" to their environments? If traits must be passed from parents to offspring, but traits *develop* as a result of gene-environment interactions, then it cannot be enough to inherit *only* our parents' genes. Instead, *parents must transfer to their offspring both their genes* and *crucial aspects of their environments.* Parental traits caused originally by gene-environment co-actions can be reliably passed to offspring *only* if parents transmit to their offspring both the genes *and* the environmental factors required for the construction of the traits. This means that constructing a comprehensive theory of biology—one that accommodates developmental, molecular (that is, genetic), *and* evolutionary data—will require an understanding of how both genetic *and* nongenetic factors are passed across generations. Thus, "an extended notion of inheritance is . . . a critical part of [developmental systems] theory." It is here, in the inescapable—but somewhat disconcerting—conclusion that we "inherit" aspects of our environments that we will find the key to reintegrating development, genetics, and evolution. But how are we to make sense of the idea that parents bequeath environmental factors to their children just as surely as they bequeath their DNA?

12

LEGACIES
How We Get Our Environments

I was just a toddler in the suburbs of Cleveland in the early 1960s, and I could not have known that a mere 10 miles away a graduate student was beginning research that, 40 years later, would help me think about the transmission of nongenetic factors from parents to their offspring. While working in the Behavior Genetics Laboratory at Western Reserve University, Robert Ressler had become interested in the effects on newborn mice of being "handled"—that is, being carried or licked—by their parents. Specifically, he sought to examine these effects because of the frequency with which claims were being made about the genetic basis of some behavioral traits commonly seen in certain strains of mice. Because the strains with which Ressler was working were genetically pure—all of the animals in a given strain shared the same genes—differences among strains that were reared in identical environments were thought to reflect their different genes. For example, it was known in the early 1960s that Ressler's white mice grew up to be heavier and more exploratory than his black mice, even when both types of mice were raised in the same highly controlled laboratory environment. Thus, it seemed that weight and exploration in these mice had to have been genetically determined.

Unfortunately, just because the logic of a study *seems* airtight does not mean it is. To see if the traits in question really could be thought of as "genetic," Ressler cross-fostered a group of mouse pups just after their birth—he placed some of his newborn white mice with black "foster" parents (the other white mice were placed with white "foster" parents), and he placed some of his black mice with white "foster" parents (again, some of his black mice were placed with black "foster" parents). He discovered that the amount of handling received by each pup varied, in part, as a function of the *parents* doing the handling; independent of the strain of the offspring, the white parents handled their adopted pups more than the black parents handled theirs. Thus, the "equal environments" that the behavior geneticists thought they were giving their genetically pure mice were not, in fact, equal. And while such a discovery is important, it pales in comparison to what Ressler discovered when he looked at the traits of the offspring later in their lives. Regardless of their genetic status, black *or* white mice who had been reared by white foster parents were heavier and more exploratory as adults than were black or white mice who had been reared by black foster parents. Clearly, the trait differences Ressler saw in the now-grown cross-fostered offspring might have resulted from the differential treatment they received as pups.

If it ended here, Ressler's tale would merely be about the dangers of concluding that genes determine traits. Instead, Ressler's next studies revealed how aspects of the developmental environment can be passed from generation to generation. In these studies, white brother and sister mice who had been reared by either black foster parents, white foster parents, or their own natural (white) parents were allowed to mate and then rear their own offspring, who, of course, were also white. A good experimentalist, Ressler implemented the parallel conditions among the black strains: black siblings who had been reared by white foster parents, black foster parents, or their own natural (black) parents were also allowed to mate and rear their own (black) offspring. His results indicated that even though all of the new generation's pups were reared by their natural parents, "both strains [of pups] . . . responded at a consistently higher rate if their *parents* had been raised by [white] rather than [black] foster grandparents." Thus, it appears that the difference in environments provided by black and white foster *grand*parents "is at least partially replicated in the parental environment subsequently provided" by *their* offspring, and that this difference then affects the behavior of the "grandpups." Michel and Moore offered the following interpretation of Ressler's important results:

> The differences in parental environment provided by the black and white foster parents so affected the development of the young they reared that, when they became mothers, they provided different environments for their own offspring than would have been the case if their mother had come from the same strain. When members of the new generation in turn became adults, characteristics of their . . . exploration were more like those of the foster grandparents' strain than like those of their own genetic strain!

Ressler concluded by noting that "a nongenetic system of inheritance based upon the transmission of parental influences is potentially available to all mammals."*

A similar nongenetic system of inheritance appears to underlie the emergence in birds of their species-typical songs, nesting sites, and nesting materials. For example, K. Immelmann found that raising two different finch species—Australian zebra finches and Bengalese finches—in each others' nests produced offspring who sang their adoptive parents' song flawlessly and exclusively. In fact, exposure later in life to their own species' song had no effect on these birds' vocalizations: they continued throughout their lives to sing the songs of the birds that had raised them as juveniles. Clearly, the same processes that led to this outcome would ensure that offspring of these birds would themselves also sing the song of their parents' adoptive parents. Findings like this led Michel and Moore to conclude that because "male bullfinches acquire the song of their fathers during their nestling stage, even if they are fostered-reared by [canary fathers] . . . there can be a line of bullfinches that sing canary songs."

Intergenerational transmission of developmentally important non-genetic factors occurs throughout the animal kingdom, sometimes in ways that do not require any sort of obvious "teaching" on the part of the parents. For example, consider the transmission across generations of non-genetic determinants of development in a species of fire ant (*Solenopsis invicta*). This species of ant is found in nature to have two distinct forms of social organization: one form has only one queen per colony, and the other has multiple queens per colony. L. Keller and K. G. Ross reported that all immature fire ant queens have similar bodies but that *mature*, "winged queens of the two social forms differ dramatically" in both their behavior and in the appearance and function of their bodies.

To study how these differences emerge in development, Keller and Ross conducted a cross-fostering experiment similar to those conducted by Ressler. They discovered that "the different maturation processes of queens of the two forms are induced largely by the type of colony in which a queen matures"; in particular, it seems that "the differences [in the maturation processes] . . . are likely to be influenced by the level of queen-produced pheromone." By itself, of course, these findings are not especially surprising; the two social forms appear to be genetically indistinguishable, so the different maturation processes would have to reflect differences in developmental environments. However, a closer look reveals a very interesting and important process by which events occurring during the development of a particular queen can have consequences that are felt by her descendants long after she is dead.

Keller and Ross write that the trait differences

> that develop between queens of the two social forms . . . influence the future reproductive opportunities of these queens, constraining them to live in a colony with the same social organization as that of the colony in which they were produced. Winged queens produced in [colonies with multiple queens] apparently have insufficient energy reserves to successfully embark on independent colony founding, which implies that they must return to an established [multiple-queen] nest in order

to survive and reproduce. . . . In contrast, winged queens produced in [single-queen] colonies acquire extensive energy reserves, which are essential for successful independent founding of colonies in ants.

And while this last observation might suggest that queens raised in the absence of sister queens get to *choose* the social form of the new colonies they start, in fact, such queens are not accepted into multiple-queen nests; consequently, queens raised as the only queen in their colony always reproduce such colonies themselves. Thus, "differences between the social forms . . . constrain the reproductive options of queens, so that the characteristic social organization of a colony is perpetuated by virtue of the social environment in which new queens are reared." While there is no obvious "teaching" going on here, it is clear that nongenetic factors in force during the development of these insects have profound effects, both on the characteristics of the developing ants themselves and on all future generations of ants that they ultimately bring forth.

Generalizing from these sorts of findings, K. Sterelny and P. E. Griffiths concluded that this type of transgenerational flow of "information" is likely to occur "in any species in which learning, broadly conceived, is important." Specifically, they state:

> Parents structure the learning environment of their young and provide them with information just through their normal, species-specific activities of daily life. So "cultural transmission" in this sense is not restricted to cognitively fancy animals.

Instead, *any* time young animals normally receive parental care over extended periods of time, there is the possibility that they will acquire characteristics like those originally acquired by their parents.

This observation has profoundly important implications, because the theory of evolution as Darwin originally conceived it requires only that *some* mechanism of inheritance ensure that adaptive traits be reproduced reliably in succeeding generations. In a classical Darwinian sense, then, some finch birdsongs, like the other traits discussed above, can be thought of as being "inheritable," even if their reliable appearance in successive generations requires nongenetic (as well as genetic) factors for their development. But Johannsen and his neo-Darwinian followers insisted that only genes can be inherited! How can nature ensure that *nongenetic* factors needed for trait development are available to offspring in successive generations?

INHERITING THE WIND: THE NATURAL SELECTION OF GENE-ENVIRONMENT SYSTEMS

Transmission of nongenetic factors across generations can occur in a couple of different ways, depending on the source of the factor. In considering this question, Griffiths and Gray found it useful to distinguish between three types of "resources" that are used by organisms as they develop.* The first type they identified, "parental resources," consists of those factors in a developing organism's prenatal or postnatal environment that are created in or by the parents and that allow development to proceed safely; resources of this sort include factors like mother-produced milk or a father-built shelter.* Clearly, when parents themselves produce nongenetic resources that offspring need to develop an adaptive trait, transmission is not a problem: parents simply provide their offspring with these factors much as they provide their offspring with their genes. The *mechanisms* by which parents provide offspring with genetic and nongenetic resources surely differ, but we need not be concerned with this difference as long as parents reliably provide both types of resources. To the extent that a father creates a developmental environment for his daughter that is similar to the environment that his father created for him, there is a sense in which aspects of the environment have been handed down through the generations.

Griffiths and Gray's second type of resource, "population-generated resources," consists of those factors that are created by the population of which the parents are a part. Communication systems and other cultural endowments, for example, are resources of this sort; a canary song in a canary's developmental environment is a population-generated resource. These resources, too, are easily transmitted across generations: parents pass on population-generated resources to their offspring just as they pass on parental resources.

The third type of developmental resource that Griffiths and Gray identified, "persistent resources," consists of those factors that persist without reference to the first two types of factors; examples of this type of resource include trees, water, and the sun. At first glance, these sorts of factors seem to be "just there"—they don't seem to be "provided" by one generation for the next. After all, parents don't produce the sun; how, then, can they "hand it down" to their offspring? Developmentalists' understanding of how these factors are transmitted across generations grows from their grasp of the workings of natural selection.

Remember: because traits are *constructed* by gene-environment coactions during an individual's lifetime, what *must* be made available to offspring if they are to develop the adaptive traits of their ancestors are *both* the genetic and environmental factors that led the ancestors to develop the traits in the first place. Only in this way can the offspring develop the adaptive traits themselves. Given this state of affairs, *Darwinian evolution can occur only when nature "selects" for reproduction in the next generation the complete* gene-environment systems *that produce adaptive traits.** After all, it is *only* in such reproduced systems that we can be confident that we will see development of the traits that so effectively served the reproducing parents. As a result, there is a fundamental similarity between developmentally necessary genetic factors and *all* types of developmentally necessary nongenetic factors; differences in the *mechanisms* by which successive generations gain access to these resources can, for the present purposes, be ignored. The upshot of this conclusion is that even persistent environmental resources such as habitat—no less than genes, or than environmental factors like parental behavior, for example—can be acquired through evolution.

To make this a bit more concrete, imagine a pair of seeds containing the same DNA, floating on the wind. Imagine further that the first seed lands on an arid island, whereas the second seed lands in a puddle in a tropical rainforest. It is entirely possible that the first seed would not develop while the second seed was thriving and multiplying. The second seed's offspring certainly inherit the second seed's DNA, but in addition, they will all develop in the environment that supported their parent's fruition in the first place. What nature has selected in this simple two-seed world is *not* a particular collection of genes alone, but rather a collection of genes *and a particular environment with which those genes can co-construct adaptive traits*. To the extent that we cannot help but develop in environments that are similar in important ways to the environments in which our parents developed, the legacy we receive from our parents includes both our genes and aspects of our developmental environments.

I have discussed with many people the idea that nature, by selecting gene-environment systems, provides offspring with the genetic *and* nongenetic factors that contributed to their parents' reproductive success; most find it intuitively reasonable. A fish out of water cannot reproduce, so all viable fish offspring develop in water, as did their parents. There is a sense, then, in which people's terrestrial environments are "given" to them by their parents just as surely as fish are "given" their watery ones. The *mechanism* by which our parents bequeath to us persistent environmental factors like air, sunlight, and gravity is certainly different from the mechanism by which they hand down their DNA, but this difference is inconsequential when we remain cognizant of the fact that natural selection operates on gene-environment *systems*. As Darwin understood in the 1850s, evolution by natural selection works *no matter how* adaptive traits are reliably reproduced in descendant generations, as long as they are.

BATTERIES INCLUDED!

One upshot of the suggestion that natural selection operates on gene-environment systems is that the nongenetic factors required for the development of evolved traits—no less than the genetic factors required for the development of these traits—can be *expected* to be available to the offspring of adapted parents. In fact, people—but also babies, embryos, zygotes, and even genes—can normally count on their local surroundings to contain the factors they need to develop evolutionarily adaptive traits; nature takes care of this for us, by selecting for survival and reproduction only those gene-environment systems that are viable.

Consider, for example, the factors immediately surrounding the genetic material. It quickly becomes obvious when considering the genes' micro-environment that we *must* inherit nongenetic factors at conception. This is because a mass of genes resting isolated in a bowl would lie there, inert, for eons (kind of like the seed that landed on the desert island);* only in the *context* of "the cellular machinery necessary for their functioning" will the genes *do* anything at all. Daniel C. Dennett refers metaphorically to this nongenetic cellular machinery as a "reader" that interprets DNA sequences:

> [W]ithout readers, DNA sequences don't *specify* anything at all—not blue eyes or wings or anything else. . . . The immediate effect of the "reading" of DNA . . . is the fabrication of many different proteins out of amino acids (which have to be on hand in the vicinity, of course, ready to be linked together). . . . So, for a DNA sequence to specify what it is supposed to specify, there must be an elaborate reader-constructor, well stocked with amino-acid building blocks. . . . This is a process that is only partly controlled by the DNA, which in effect *presupposes* (and hence does not itself *specify*) the reader.*

The presence in the genes' environment of the required biochemical machinery is, of course, virtually guaranteed, because nature selects only viable gene-environment systems, favoring genes that *come with* their own machinery. In fact, *a zygote inherits from its parents more than just its genes;* it also inherits all of the *non*genetic materials that constitute the bulk of its mass, including the egg's membrane, its cytoplasm, and all of the cytoplasm's contents. Among these contents are maternal cellular "organs" (called "mitochondria") that provide the zygote with needed energy, and factors such as maternal RNA products that are nonrandomly distributed throughout the cytoplasm. These nongenetic—but inherited!—factors play important roles in determining aspects of an animal's basic body structure, including which end of the zygote will become the animal's head.

Evidence that such nongenetic factors are literally inherited was provided in 1983 by M. W. Ho, C. Tucker, C. Keeley, and P. T. Saunders. These researchers knew initially that exposing a fruit fly embryo to ether disrupts the development of a normal fruit fly body plan. What they discovered was that the abnormal body plan typical of ether-exposed flies could be *inherited* via ether's effects on the *cytoplasm* of an egg cell. It turns out that changes in any of the *non*genetic materials that contribute to a fertilized egg "can cause heritable variation that appears in all the cells descended from that egg cell." Thus, it is here, at the very conception of a life, that we can see most easily how developmentally impoverished the neo-Darwinian synthesis really is; we simply do not inherit *only* our parents genes, as that portrayal suggests. Instead, as Ho points out, "The material link between . . . development and evolution . . . is the hereditary apparatus, which realistically includes both cytoplasm and nuclear genes."

We can easily extend this analysis to the level of the environment slightly farther removed from the genes: the immediate surroundings of the zygote. Much as a mass of genes in a dish cannot *do* anything at all, a newly conceived embryo located in an inhospitable environment—for example, in a uterus unprepared for pregnancy—will not develop. Thus, an embryo cannot develop *even if it possesses a full complement of "normal" genes* and *all of the cellular machinery required for "reading" DNA*. In the same way that inheriting genes is useless unless we also inherit the cellular machinery needed to "read" DNA, so inheriting both genes *and* cellular machinery is useless unless embyros—and later, babies and children—have access to environmental factors needed to support their survival. This same logic can be extended into each of the other levels of the environment surrounding the genes: because natural selection operates on gene-environment systems, the *macro*environmental factors required for the development of adaptive traits—like the requisite genetic factors—are normally available to developing organisms. Furthermore, this analysis holds for each stage through which developing organisms pass; consequently, environmental factors required for the development of evolved traits *in adolescence* are normally available to developing adolescents.

THE FUNDAMENTAL SIMILARITY OF A CABIN AND A CAVE

Notwithstanding developmentalists' willingness to ignore the fact that genetic and nongenetic factors become available to descendants via different mechanisms of intergenerational transmission, do they distinguish between different sorts of *nongenetic* developmental resources? After all, persistent resources like oxygen, water, and moderate temperatures are available to descendants in a decidedly passive way, but population-generated resources like language are available only if actively provided by a care-giving adult.

After looking closely at the differences between different types of resources, Griffiths and Gray concluded that, for our purposes, sharp distinctions between the different types of factors are not warranted. Specifically, they observed that "there is a fundamental similarity between building a nest [a parental resource], maintaining one built by an earlier generation [a population-generated resource], and occupying a habitat in which nests simply occur (for example, as holes in trees) [a persistent resource]." In each case—regardless of the *origin* of the "home" in which a newborn animal finds itself—the developmental environment is specific to the organism, it is somewhat organized and predictable, and it limits the stimuli that can reach its young occupants. As a result, *each* of the different types of resources can and should be seen "as a provider of basic, necessary elements for development and as a *positive, informative,* and *constructive* force" in trait development. Thus, developmentalists think that distinctions between different resource types are unnecessary. The fact that various features of the environment become available to developing organisms in different ways is as inconsequential as the fact that genetic and nongenetic factors are bequeathed to us via different mechanisms. It doesn't matter to evolutionary *or* developmental processes *how* a developing animal's environment comes to be how it is; as long as complete gene-environment systems can be reproduced in successive generations, evolution will proceed as Darwin reckoned it would, and traits will emerge as developmental biologists have discovered they do.

ORGANISM-ENVIRONMENT CO-EVOLUTION

Another upshot of the suggestion that natural selection operates on gene-environment systems—one that I find really amazing—is that the adaptive "fit" of a species to its environment need not be achieved only by nature "selecting" those organisms that are adapted to their environments. Instead, a good "fit" can also arise as a consequence of behaviors that cause *environments* to be adapted to the behaving organisms. This argument rests on the well-known tendency of animals to *construct* aspects of their environments, as happens when birds build nests, beavers build dams, and bees build hives. Given the ubiquity of this phenomenon in the animal world, Lewontin was emboldened to write:

> The final step in the integration of developmental biology into evolution is to incorporate the organism as itself a *cause* of its own development, as a mediating mechanism by which external and internal factors influence its future. . . . The organism is not simply the object of developmental forces, but is the subject of these forces as well.

Thus, while it is *possible* that some organism-environment compatibilities result from natural selection operating on organisms and that others result from organisms operating on their environments, the fact that animals are subjects *and* objects of developmental forces means that in all likelihood, "the adaptive fit between organisms and their environments could be caused by both of these processes acting together."

In an extraordinarily insightful chapter presenting his aptly named theory—the theory of *organism-environment co-evolution*—F. J. Odling-Smee offers the following illustration of phenomena that might be best understood in this way:

> The Galapagos woodpecker finch . . . is famous for using a tool—a twig or a cactus spine—which it holds in its beak to poke out grubs from the barks of trees. . . . Probably its novel behavior is based on learning rather than on gene mutations. . . . One apparent result is that these birds have never evolved either the typical beak or the typical tongue of other woodpeckers. The immediate question is Why not? This expands to a more interesting question: What exactly is being transmitted via evolutionary descent among successive generations of woodpeckers? Part of the answer is clear enough. If Alcock is right, a capacity for learning the tool-using skill must be transmitted via genes in the usual way. But that alone is not enough to account for the whole phenomenon. It is also likely that the particular "woodpecking niche" opened up for these birds by their own behavior is transmitted from generation to generation, not genetically, but via those environments which successive generations of woodpeckers are themselves ensuring they encounter. In that case, a transmitted niche could then explain the modification of the selection forces in the finches' environment, which, on the one hand, are apparently selecting for tool-using skills but which, on the other hand, are clearly not selecting for the elongated beaks and tongues of typical woodpeckers.

Given intergenerational transmission of the skills required for survival in these finches' niche, it is unlikely that these birds would ever develop beaks typical of woodpeckers; the selective pressure to do so is simply not there. Instead, these birds have co-evolved with aspects of their environment—in particular, with those stimuli that allow finches to learn the tool-using skills needed for grub extraction—such that they are now thriving in their niche even in the absence of evolved anatomical adaptations like woodpeckers' beaks.

Thus, far from merely "supporting" the activities of the genes, elements of the environment—from factors in a fetus's environment (for example, sounds that influence preferences at birth), to persistent factors in a newborn's environment (for example, light that, when detected binocularly, contributes to the development of normal brain structures), to parentally provided factors in a juvenile's environment (such as specific languages that determine how we communicate)—all of these elements play active roles in giving specific *shapes* to behavioral characteristics. And to the extent that these characteristics contribute to our survival, we will have the opportunity to bequeath to our children those portions of our DNA *and* those environmental factors that, together, produced these characteristics in us in the first place. In this way, a child inherits both "a structured genome and a structured segment of the world . . . not just genes, but a host of other necessary influences and interactants as well."*

The conclusion that "the full range of developmental resources represents a complex system that is replicated" during the development of successive generations is rather startling, given how narrowly heredity has been understood since the emergence of the modern synthesis. Nonetheless, further consideration bears out the value of adopting such "an extended notion of inheritance." After all, "elements of culture, such as . . . social structures . . . are required for the replication of evolved psychological traits [such as language] in humans."

But if "necessary influences and interactants" are transmitted from generation to generation, then any effects they have on our traits—*even if the effects are manifested during our lifetimes*—can conceivably be "passed on" to our offspring. Does this mean that Darwin's conviction that acquired characteristics can be inherited was well founded in the first place?

13

On Big Muscles and Facial Hair

Reconsidering "Inherited," "Acquired," and "Innate"

By the time I found my fabulous dog, James, he was already a full-grown adult. It was clear that he had been on the streets for a while, as his long hair was matted and he was very thin. Having never had a dog before, I didn't know what to do with him, but my girlfriend at the time thought we should take him home for a good meal and see if we could track down his owner.

When I woke up the next morning, James was sitting by the door, quietly watching us sleep. Even to a dog-ignoramus like myself, it was clear that he needed to go out; apparently, he had already been house-trained. Later, when I took him to school with me—I was afraid to leave him in my house alone, not knowing what sorts of mischief he might get into—he sat in the classroom while I lectured, listening attentively.

That was ten years ago. Weeks of searching never turned up our abandoned dog's original owner, but the days together did generate the kind of dog love that prevents following through on one's plans to take a foundling to the pound. In the ensuing months, it became clear that James would *never* have chewed on any of my belongings had I left him alone in my house, but it was quite some time before my hunch was put to the test, because our dog-in-the-classroom routine was so good for everyone involved that he wound up accompanying me to work every day. While my students loved him, he was particularly appreciated in faculty meetings for his patient, benevolent, and noble demeanor.

Shortly after I found James, a few questions began to surface in my mind. Would James have been a nonchewer as an adult if he had spent his puppyhood with me, or did this characteristic, like the fact that he had arrived in my life already house-trained, reflect the specific experiences he had had with his original owner? Was James's calm disposition a function of these early experiences, too? Rephrasing this question, was James's relaxed temperament "inherited" from his biological parents or "acquired" as a result of particular experiences he had as a puppy? And if he had ever had puppies himself, would they have "inherited" any of the traits he acquired as a result of his life experiences? Some preliminary discussion is needed before we can address these questions with any clarity; the idea that traits can be "inherited" or "acquired"—or, in some cases, "acquired" in one generation and then "inherited" by the next—is very old, so these words come with numerous connotations that should be identified before we proceed.

AN ACQUIRED BEARD

The notion that "acquired" characteristics might be "inheritable" has a long and checkered past, beginning—for our purposes—in 1809, with the publication of Jean Baptiste de Lamarck's masterpiece, *Zoological Philosophy: An Exposition with Regard to the Natural History of Animals*. Lamarck, a French botanist and zoologist who coined the word "biology" in 1802, used the word "acquire" to mean "gain as a result of effort." In contrast, he wrote as if a trait was "preserved by reproduction" whenever a descendant's trait resembled a trait found in the descendant's ancestor. Thus, Lamarck believed that an *acquired* characteristic was "inherited" whenever a trait that appeared because of an animal's *efforts* subsequently appeared in that animal's descendants (for whatever reason).* This definition made sense given Lamarck's goal: to make a plausible case for his then-novel idea that species *evolve* through modifications of traits that are subsequently passed from generation to generation.* Since Lamarck was interested in making the case that species evolve, what concerned him was that acquired characteristics could be "preserved" from generation to generation; he was not concerned with *how* that could happen. Similarly, Darwin wrote as if he believed that ancestral traits that are reliably reproduced in descendants have been "inherited" by those descendants.

By the end of the 1800s, August Weismann—of tail-chopping fame— was using the phrase "inheritance of acquired characteristics" much more narrowly. Weismann maintained that an acquired trait could be counted as "inherited" only if the ancestor's trait itself "(and *not* the environmental agent which affected it) induces a change in the [ancestor's] germ plasm which in turn produces the same . . . modification in the following generation." Thus, Weismann would have conceded that descendant mice had inherited the tailless characteristic of their ancestors only if chopping off the ancestors' tails somehow affected the ancestors' "germ plasm" so that their sperm and egg cells now carried "determinants" for tailless offspring.

Evidently, in contrast to Lamarck, Weismann cared a lot about *the mechanism* by which acquired characteristics might be inherited.

Many people today are comfortable labeling traits "inherited" only if they develop *independently* of specific experiences. For example, we usually consider facial hair to be inherited in men, because it emerges on an adolescent's chin in a way that appears to be independent of the adolescent's experiences. In contrast, many people would call a trait acquired only if it develops in response to specific experiences. For example, most of us would consider massive muscles to be acquired, because they develop only after we lift heavy weights. Given these understandings, then, an acquired characteristic can be inherited only if a trait that appears in one generation *because of specific experiences* can subsequently appear in later generations *before the individuals of that generation ever have those experiences.* Thus, while Lamarck was unconcerned with the circumstances under which traits might appear in successive generations, many modern biologists—and much of the twenty-first-century public, too—think that such things are quite important.

Clearly, scientists of the last two centuries have given us varying ways to conceptualize "the inheritance of acquired characteristics"; as a result, we should not be surprised to find conflicting claims about the existence of this phenomenon. For example, Weismann's intellectual descendants remain confident that acquired characteristics *cannot* be inherited, because there is currently little evidence that one's experiences can actually change one's DNA in a way that can be passed on to one's offspring.* The same skepticism characterizes those who maintain that traits are "inherited" only if they develop in the absence of specific experiences; after all, a weight lifter's daughter obviously won't develop large muscles in the absence of the weight-lifting experiences that produced her parent's physique.

In contrast, using a more Darwinian conceptualization of "inheritance"—a trait is inherited whenever it is reliably reproduced in successive generations—developmentalists sidestep questions about the inheritance of acquired characteristics completely. In fact, developmental systems theorists see no line between inherited and acquired traits at all. Upon further reflection, this should not be surprising. The commonly accepted inherited/acquired dichotomy reflects the idea that inherited genetic factors cause inherited traits and that experiential factors cause acquired traits. But developmentalists find neither of these factors causing trait development without the help of the other. If developmentalists find the genetic/nongenetic dichotomy to be a false one, it should be no wonder that they find the inherited/acquired dichotomy to be a false one as well. This was the point Griffiths and Gray made when they argued that the traditional—and spurious—distinction between biological and "cultural" evolution "rests on a distinction between genetically 'inherited' and environmentally acquired traits":

> The developmental systems view implies that it is not possible to divide the traits of organisms into those with a genetic base, which can be explained by biological evolution, and those which are environmentally acquired and are the domain of cultural evolution. The means by which traits are reconstructed in the next generation are varied, and do not admit of any [such] simple twofold division.

A concrete example will be clarifying here; let's further consider the development of facial hair. Facial hair seems to develop independently of experience, and so seems to be "inherited." Nonetheless, given the developmental tenet that Johannsen gave us a century ago—traits themselves cannot be inherited but rather must be *constructed* during development—*all* of our traits must be *acquired* during our lifetimes. Think of it like this: you first existed as a zygote, and since you shared *no* observable characteristics with other recognizable members of our species when you were in this form, *every one* of your currently observable traits has to have been "acquired" over the course of your lifetime. The emergence of facial hair, like the emergence of any trait, depends on more than just inherited genetic factors (remember, without the requisite hormonal milieu—which *can* be completely independent of genetic factors, as it is in the case of eunuchs,* for example—no beard will grow). Thus, as developmentalists see it, boys must *acquire* facial hair as they develop, just as they must acquire *all* of their traits.

By the same token, developmentalists see the converse as being true as well: *all* traits, including those that are acquired only after specific experiences, have the potential to be "transmitted" to—that is, reproduced in—offspring. According to this perspective, if the processes that give rise to a trait during a lifetime show enough "stability over evolutionary time" that the trait consistently develops in each new generation, the trait will be subjected to natural selection and qualify as potentially "transmittable." In this way, *all* traits have the potential to be "transmitted" from generation to generation; such transmission merely requires offspring to have both the DNA and the developmentally important experiences that their parents had as they developed. Therefore, evolutionary processes can lead to the reproduction of traits in descendant generations regardless of the extent to which the traits depend on specific experiences for their development.

What this means is that all traits have the potential to be "transmitted" to offspring, just as all traits are acquired. This insight led the great developmental biologist Conrad Waddington to write:

> The exclusion of acquired characters from all part in the evolutionary process does less than justice to the incontrovertible fact that they exhibit

some of the hereditary potentialities of the organism. All characters of all organisms are, after all, to some extent acquired characters, in the sense that the environment has played some role during their development. Similarly, all characters of all organisms are to some extent hereditary, in the sense that they are expressions of some of the potentialities with which the organism is endowed by its genetic constitution.

But if *all* traits are acquired and *all* traits are transmissible across generations, then it is not possible to distinguish between "inherited" and "acquired" traits. Thus, Gottlieb was on firm ground when he wrote, "There is no basis for the dichotomy of heritable-nonheritable because all traits arise as a consequence of . . . development."

Adopting a developmental systems perspective, then, effectively short-circuits the 200-year-old debate in biology about the inheritance of acquired characteristics. *Certainly, there are valuable distinctions to be made between the emergence of traits like large biceps and the emergence of traits like facial hair;* I will address this issue shortly. But in both cases, the genetic *and* nongenetic factors required for the construction of the trait *can* be passed from parents to offspring, and the offspring can then *construct* the trait *during their development* just as their evolutionarily successful parents did before them. This process characterizes how *all* traits are reproduced in successive generations, regardless of whether we are talking about traits like facial hair, which do not seem to require experience for their appearance, or traits like toned muscles, which do.

Note that the ways of thinking favored by developmentalists are consonant with the essential Darwinian proposal that evolution will occur in any world with limited resources, provided that ancestors' adaptive traits are reliably reproduced in their descendants. What changes when a developmental systems perspective is adopted is the idea that natural selection operates on *genes alone*, an idea that emerged from the modern synthesis, not from Darwin. Darwin's belief that natural selection operates on *traits* is compatible with the developmental systems tenet that natural selection operates on gene-environment systems, since traits emerge from such systems.*

TRAIT MALLEABILITY
AND ENVIRONMENTAL IMPERVIOUSNESS

The notion that *all* traits are acquired and that *all* traits have the potential to be transmitted across generations might leave some readers with the uncomfortable sense that developmentalists cannot distinguish between traits that really ought to be distinguished. Does dispensing with the inherited/acquired, nature/nurture, and gene/environment dichotomies* leave us powerless to distinguish between traits like well-developed muscles that obviously require specific experiences for their appearance, and traits like two eyes that develop in ways that seem impervious to experience? After all, the emergence of some of our traits depends very much on who raises us and how, but other traits seem to emerge independently of such factors.

In fact, developmentalists do acknowledge the value of this distinction (of course). Timothy D. Johnston, whose 1987 critique of dichotomous thinking served as my introduction to the developmental systems perspective, wrote: "It is clear that some behavioral patterns are highly resistant to experiential modification and others are less so; that some patterns are present at birth and others are not; that some patterns depend on practice for their normal development whereas others require less obvious kinds of experience." *What separates those who think developmentally from those who do not is the understanding that traits that seem impervious to experience are no more "genetic" than are traits that seem "open" to such influence.* The extent to which experiences influence a trait's development reflects a variety of factors (some of which I will consider below), *but it does* not *reflect the extent to which genes control the trait's development.*

Traits that seem to appear independently of developmental circumstances are often said to be "innate." To get a clear understanding of how developmentalists distinguish between such traits and traits that seem "open" to environmental influences, it will be useful to consider the concept of "innateness" in some detail.

Innateness

The word "innate" has its roots in the Latin word *innasci*, meaning "to be born"; the first definition (1a) of "innate" offered by *Webster's Third New International Dictionary* is "existing in or belonging to some person or other living organism from birth." Accordingly, physical traits like four limbs in mammals or behavioral traits like human sucking reflexes are considered "innate" because they are present at birth. *Webster's* also defines "innate" (definition 1c) as meaning "inherent . . . rather than derived from experience"; this broader definition derives from the fact that newborns' traits *appear* to develop in the absence of experiences. Some traits, then—including human physical traits like facial hair or behavioral traits like walking—are considered "innate" even if they are not present at birth, as long as they seem to develop independently of an organism's experiences. In addition, once present, "innate" traits often appear to be unchangeable by environmental manipulations that are anything short of catastrophic. Thus, in common usage, the word "innate" comprises a variety of subtly different notions, including "early-appearing," "unlearned," "impervious to environmental influence," "inevitable" (referring to "certainty of occurrence"), and "immutable" (referring to "nonsusceptibility to influence once present"). Can *Webster's* definitions help us distinguish between traits that seem "open" to environmental influences and those that seem impervious to such influences?

Because studies of the effects of prenatal exposure to sensory stimuli indicate that the moment of birth is, in many respects, rather arbitrary, *Webster's* first definition of "innate"—which hangs on this moment—cannot yield the distinction we seek. Generally, scientists who study perceptual and behavioral processes in newborns agree that while the experience of traveling through a birth canal is undoubtedly novel and stressful, a person's birthday is really just another day in the continuous series of days that stretch from conception to death. Presumably, the day of birth *seems* like a pivotal moment in a baby's life because many of us are under the impression that experience does not begin until then. However, while it is true that some stimuli are unavailable to fetuses, it is simply not true that fetuses have no experiences; you will recall that sounds heard by human fetuses affect their postnatal auditory preferences.* Instead, *every* trait has a developmental history—traits do not appear out of the blue, fully formed like Athena from the head of Zeus—and traits always emerge in *some* potentially influential environment (even if that environment is a uterus). These observations led Johnston to remark that although some traits are present at birth and others are not, we must nevertheless resist "the temptation to see these classifications as implying different kinds of explanation for development. . . . Pointing out that a pattern of behavior [or, for that matter, a physical trait] is present at birth does not explain *how* it comes to be present at birth—the problem of development needs to be solved for congenital as well as [later-developing traits]."

Because experiences can affect the traits of *both* adults and newborns, and because some traits seem to emerge independently of experience *regardless of when* in development they appear, it is a mistake to maintain a qualitative distinction between early-appearing and later-appearing traits. Focusing on whether or not a trait is present at birth will not help us distinguish between traits that are "open" to environmental influences and those that seem impervious to such influences.

Webster's second definition of "innate" is only slightly more useful than its first definition. Defining a word negatively—*not* derived from experience—is of limited help; if a trait is not derived from experience, what *is* it derived from? In many cases, when people label a trait "innate," they seem to be thinking that the trait is somehow specified "in the genes." Those unfamiliar with the epigenetic nature of development sometimes imagine that our traits are constructed by two distinct developmental mechanisms, one that builds innate traits using only genetic information, and another that builds noninnate traits by also utilizing environmental input. Unfortunately, such a portrayal is not in accord with what is known about how traits really develop. Given that all traits develop through gene-environment co-actions, what exactly is going on when a trait seems to develop independently of environmental factors?

Canalization

Waddington introduced the developmental concept of "canalization" specifically to address this question. He defined canalization as "the capacity to produce a particular definite end-result in spite of a certain variability both in the initial situation from which development starts and in the conditions met with during its course." To help people understand this idea, he asked them to picture development as being analogous to what happens when a ball is placed at the top of a mountain; when an organism begins to develop, it starts heading down a developmental pathway just as a ball placed on a mountaintop begins rolling downhill. Along the pathway are many branching points, each of which can carry the organism to a different endpoint. In 1974, Waddington portrayed a trait's resistance to change as "a set of branching valleys," writing that trait development

> proceeds along a valley bottom; variations of genotype, or . . .
> environment, may push the course of development away from the
> valley floor, up the neighbouring hillside, but there will be a tendency
> for the process to find its way back, not to the point from which it was
> displaced . . . but to some later position on the "canalized pathway of
> change" from which it was diverted. Or, if it is pushed over a watershed,
> it may find itself running down to the bottom of another valley.

Thus, Waddington's so-called epigenetic landscape allows us to see, metaphorically, that certain developmental outcomes are more likely than others, sometimes regardless of the minor variations in "terrain" encountered by different organisms as they wend their way through their unique lives. "Once a pathway is chosen it is entrenched or bound to produce a particular end state. It is this entrenchment that Waddington called canalization." Such an arrangement ensures that only unusual environmental or genetic disturbances prevent the development of certain characteristics.

The concept of canalization, then, allows us to distinguish traits that seem to develop independently of environmental factors and traits that seem to be sensitive to such factors. Specifically, highly canalized traits develop in many different environments and so appear to develop independently of environmental factors; the pathways leading to these traits can be pictured as very deep trenches that strongly discourage (but nevertheless allow in unusual circumstances) deviation from the path. In contrast, less canalized traits develop via travel down relatively shallow pathways; the emergence of these traits seems to be much more sensitive to organisms' developmental environments. But is Waddington's conceptualization just a restatement of old concepts in new words? Are highly canalized traits merely traits that are genetically determined or otherwise "innate" (in the sense of "not derived from experience")?*

Not at all. While the concepts of innateness and canalization both distinguish environmentally sensitive traits from those that seem impervious to such influences, only canalization does so in a way that is consonant with what is known about the epigenetic nature of trait development. The concept of innateness often carries with it the idea that traits that seem insensitive to environmental factors are genetically determined, an idea that is demonstrably false; in contrast, the concept of canalization does not imply anything about genetic determination, because canalization need not be controlled strictly by genetic factors. This became apparent when Gottlieb discovered a phenomenon that he has since dubbed *experiential canalization*.

Gottlieb's studies of mallard ducklings' attraction to the "assembly call" of their mothers demonstrated that their attraction depends in part on the ducklings having been exposed before hatching to embryonic mallard vocalizations. Although ducklings that hear these sounds as embryos always develop preferences for the *mallard* maternal call, mallard embryos that are *prevented* from hearing these sounds grow to be *as attracted* to other bird species' maternal calls as they are to their own. After years of thought, Gottlieb understood this finding to mean that a mallard embryo's auditory *experiences* might play a canalizing role in development; that is, some early auditory experiences might make a mallard unsusceptible to the influence of subsequent auditory stimuli. A strong test of this hypothesis, of course, is to see if early, exclusive experience with *non*mallard calls can render a mallard embryo unresponsive to later exposure to *mallard* calls. Given that normal mallards are so attracted to mallard calls that previous generations of scientists had assumed the attraction was innate, Gottlieb thought it unlikely that a mallard could ever be induced to *prefer* another species' call over its own; he wrote "To be quite frank, I did not believe it was possible to demonstrate malleability in the presence of the species-specific maternal call."

Nonetheless, Gottlieb pressed on, studying the effects of preventing mallard embryos from hearing embryonic mallard vocalizations and instead exposing them to the maternal calls of chickens or wood ducks. Amazingly, he discovered that with these manipulations, he could induce mallards to *prefer* the calls of these other species over the calls of their mallard kin. Perhaps even more important, he discovered that mallard embryos exposed to *both* the chicken call and embryonic mallard vocalizations did *not* come to prefer the chicken call. Instead, exposure to the embryonic mallard sounds apparently *buffered* development of the mallard call preference from the influence of the chicken call; consequently, mallard calls came to be preferred even in the face of significant exposure to the chicken call. Gottlieb concluded, "exposure of mallard ducklings to their [embryonic vocalizations] . . . buffers the duckling from becoming responsive to social signals from other species. In the absence of exposure to the [embryonic vocalizations], the duckling is capable of becoming attached to the maternal call of another species."

Gottlieb's work shows that while the development of the mallard call preference is highly canalized, the canalization is not exclusively controlled by the genes. Instead, an *experience* can apparently serve to shield development, canalizing it such that a particular outcome will occur in a startlingly wide range of post–experience environments—a range so wide, in fact, that at first glance, the outcome might actually appear to be insensitive to experiential factors. Since canalization can arise not only from genetic factors but also from an organism's experience (and from factors at intermediate levels of developing systems), the concept of canalization is not plagued by the problems that plague the concept of innateness (with its accompanying notion of genetic determination). Rather, the concept of canalization can help us see that although some traits are exquisitely sensitive to environmental circumstances and others seem impervious to such factors, both the traits themselves *and the components of the system that buffer trait development* emerge from nongenetic as well as from genetic factors. The concept of canalization, then, allows us to distinguish traits that seem insensitive to environmental influences from those that do not, but in a way that is compatible with the known facts of development.

Some theorists have suggested that we could tweak our understanding of "innateness" to allow it to capture the epigenetic qualities of real development. This is certainly possible, as "innate" need not *necessarily* imply "genetically determined." Nonetheless, because the notion of "innateness" currently carries this implied baggage, I prefer to abandon it entirely in favor of the notion of "canalization," which requires no tweaking at all. In justifying his decision to coin the new word "gene," Johannsen wrote:

> It is a well-established fact that language is not only our servant . . . but that it may also be our master, overpowering us by means of the notions attached to the current words. This fact is the reason why it is desirable to create a new terminology in all cases where new or revised conceptions are being developed. Old terms are mostly compromised by their application in antiquated or erroneous theories and systems, from which they carry splinters of inadequate ideas, not always harmless to the developing insight.

To avoid the "splinters of inadequate ideas" associated with the concept of innateness—namely the idea that some traits are genetically determined—

I refrain from using this concept entirely, preferring to use the concept of "canalization" in its stead. After all, the concept of canalization does everything we need it to do: it allows us to distinguish between traits like hairstyle that are obviously affected by environmental factors, and traits like facial structure that seem impervious to such factors, but it does so without encouraging us to imagine that some traits are more influenced by genes than are others.

The Illusion of Imperviousness to Environmental Influence

In the preceding paragraphs, I have always written that highly canalized traits *seem* to be relatively impervious to the influence of environmental factors, not that their development *is*, in fact, relatively impervious to such factors. This is not simply the conservative wording of a cautious academic. Rather, this wording reflects the fact that DNA cannot construct traits without environmental input. Given this fact, it *must* be that traits sometimes *seem* environmentally insensitive even when in reality, they are not. But how can such an illusion of imperviousness to environmental influence occur?

Such an illusion can occur whenever the environmental factors that contribute to a trait's appearance are unknown but ubiquitous in a species' developmental environment. In situations where such contributing factors are always present, the traits will develop *regardless of variations that might occur in any of the other noncontributing factors that characterize the environment.* For example, recall that normal mallard ducklings preferentially approach their mothers' calls regardless of the environments in which they are reared; such ducklings can be reared in the woods, in incubators, or even with chicken foster parents, and still they develop a preference for this call. Thus, development of this preference is so highly canalized that to anyone unaware of how mallard *embryos'* experiences affect its development, it *appears* impervious to environmental influence. Since the vocalizations that affect the development of this preference are available in the environments of virtually all mallard embryos, mallard hatchlings can be reared in an extraordinarily wide range of environments and still have a preference for their mothers' calls. The illusion of imperviousness to the environment, then, arises whenever the environmental factors that contribute to a trait's development have not yet been discovered but are nonetheless ubiquitous in the developmental environment.* The fact that ubiquity can render such factors effectively "invisible"—people have a hard time noticing things that never vary—underscores the need to be vigilant, so as not to be duped by this illusion.

The illusion of imperviousness to environmental influence is most likely to occur when there are relatively few macroenvironmental factors involved in a trait's development. This is because this illusion arises when unknown but influential factors are ubiquitous in the developmental environment, and the odds that *all* of a particular trait's influential factors will be ubiquitous decreases as the number of environmental factors involved in the trait's development increases. For example, relatively few macroenvironmental factors are involved in the development of disorders like PKU, which develops when someone unable to break down phenylalanine nonetheless consumes it. Since only one step on the developmental pathway leading to PKU involves the external environment—namely the one in which a person consumes phenylalanine—there is only one environmental factor that needs to be ubiquitous in order for PKU to *appear* impervious to environmental influence. And, in fact, because phenylalanine *is* ubiquitous in human developmental environments, people with the genetic defect "for" PKU develop this disease in all normal human environments. Consequently, until we understood how PKU develops, it *looked* impervious to environmental influence; only solving the riddle of its development allowed us to see that its emergence actually *depends* on the presence of a specific environmental factor, namely dietary phenylalanine.

In contrast, there are many more macroenvironmental factors involved in the development of an above-average IQ, so it is unlikely that *all* of the environmental factors that contribute to this trait are ubiquitous in human developmental environments. As a result, this trait is unlikely to appear impervious to environmental influence. And sure enough, performance on IQ tests is *obviously sensitive* to environmental factors, including prenatal and postnatal nutrition, cultural values, and access to education.

Thinking about trait development in this way can help us take our intuitive—but developmentally nonsensical—impression that traits vary in the extent to which they are "genetic," and square it with the developmentally sound idea that *all* traits are constructed by equally important contributions of genetic and nongenetic factors. Our sense that some traits are relatively impervious to environmental influences invariably reflects incomplete understanding of the ways in which genetic and environmental factors interact in the development of those traits. Thus, while traits whose development is not highly canalized (for example, hairstyles) can *obviously* be affected by environmental factors, traits whose development *is* highly canalized (for example, PKU) merely *seem* impervious to such factors; when we finally understand how genetic and nongenetic factors co-act to produce a trait, we can see how that trait can be influenced by environmental manipulations regardless of how highly canalized its development is. According to this perspective, then, an important difference between traits traditionally thought of as "acquired" and traits traditionally thought of as "inherited" is that the former develop when genes interact with known environmental factors, whereas the latter develop when genes interact with ubiquitous environmental factors that are currently unknown.

What do these conclusions imply about the origins of my dog James's temperament? Did he develop his calm nature because of his early experiences with his original owner, or would this disposition have developed even if he had been mine from birth? Because we know very little about how genetic and nongenetic factors interact to produce a dog's temperament—let alone a person's!—there cannot yet be a satisfying answer to this question. One reason we cannot get beyond generalizations in this case is that without knowing what environmental factors contribute to a dog's temperament, there is no way to know how ubiquitous these factors are. If they are present in *all* human households, the development of James's temperament would have been highly canalized, and he would have developed the same temperament in my household as he did in his original owner's. On the other hand, if the factors contributing to James's temperament varied across these two environments, his temperament would have developed differently in the two circumstances. Thus, there is currently no way to discern how James's temperament would have developed if he had spent his puppyhood with me.* Nonetheless, we *can* be confident that his temperament reflected the influence of both nongenetic *and* genetic factors, because *all* traits are acquired as a consequence of gene-environment co-actions. The developmental systems perspective does not necessarily offer answers; instead, it very often helps us see how much more research we need to do before answers will be forthcoming.

Note that the developmental systems perspective does *not* call into question the importance of either genes or of natural selection in the development or evolution of our traits; both obviously play important roles in these phenomena. Rather, this perspective merely encourages us to recognize the roles of additional factors in these processes. Furthermore, it throws into relief the problems that arise with dichotomous thinking about so-called inherited and acquired traits. In so doing, it helps us see that acquired traits can be passed from generation to generation; this conclusion follows from the realization that natural selection provides organisms with both the genetic *and* environmental factors that characterize their developmental circumstances. As soon as we acknowledge that *all* traits are acquired when genes and environments co-act during development and that natural selection gives us *both* the genetic and nongenetic factors that we need to build our traits, "Heredity . . . becomes inseparable from development." Ultimately, the developmental systems perspective gives us elbowroom that permits construction of a theory of biology that encompasses evolution, genetics, *and* development.

14

THE FETAL APE
Consequences of Integrating
Development and Evolution

In 1937, five years after stunning Western civilization with his darkly satiric prophetic novel *Brave New World*, Aldous Huxley moved from Europe to southern California. Having struggled with near-blindness for 26 years, Huxley's doctors thought such a move would be good for the health of his eyes. Two years later, he published another, less well-known satire in which he pondered man's hunger for immortality. Set primarily in late 1930s Los Angeles, *After Many a Summer Dies the Swan* tells the story of a millionaire oil tycoon who, at the age of 60, has a relatively mild stroke and subsequently develops a profound fear of death.* And in a Hollywood environment not so different from the one that lives on today, he does what some other extraordinarily rich men might do in a similar situation: he hires a personal physician—one Dr. Sigmund Obispo—and begins bankrolling the latter's research on extending the human life span.

At the end of the book, Dr. Obispo and his wealthy patient travel to Europe to search out the Fifth Earl of Gonister, who, Obispo has come to believe, unraveled the mystery of immortality hundreds of years earlier in eighteenth-century England. After a bit of searching, they find him living with his housekeeper in a foul-smelling subterranean chamber, having celebrated his two-hundred-and-first birthday a few months earlier. The secret to the Earl's longevity? Daily ingestion of raw fish entrails, carp to be specific. But it is the *end result* of 200 years of individual human development that really shocks Huxley's rich industrialist:

> On the edge of a low bed, at the centre of this world, a man was sitting, staring. . . . His legs, thickly covered with coarse reddish hair, were bare. The shirt, which was his only garment, was torn and filthy. . . . He sat hunched up, his head thrust forward and at the same time sunk between his shoulders. With one of his huge and strangely clumsy hands, he was scratching a sore place [on his leg]. . . .

"A foetal ape that's had time to grow up," Dr. Obispo managed at last to say. . . . Above the matted hair that concealed the jaws and cheeks, blue eyes stared out of cavernous sockets. There were no eyebrows; but under the dirty, wrinkled skin of the forehead, a great ridge of bone projected like a shelf. [At this point, Obispo's patron asks the doctor who this monstrosity is, and Obispo confirms that it is the two-hundred-and-one-year-old Fifth Earl of Gonister.]

Dr. Obispo went on talking. Slowing up of development rates . . . one of the mechanisms of evolution . . . the older an anthropoid, the stupider . . . no death, perhaps, except through an accident . . . but meanwhile the foetal anthropoid was able to come to maturity.

If this story grew only from the imagination of one of the twentieth century's most creative novelists, we could understand it simply as a good old-fashioned Hollywood horror story cum cautionary tale. But Aldous Huxley was no mere spinner of fantastic yarns; his stories reflect the fact that he followed advances in biology throughout his lifetime (indeed, only his failing eyesight kept him out of medical school as a young man). No doubt his older brother helped him develop his biological erudition; Sir Julian Huxley was one of the most famous biologists of the mid-twentieth century. The climax of Aldous Huxley's Hollywood novel hints at some remarkable ideas about the relationship between individual development and evolution, ideas that had only recently dawned on biologists who were trying to integrate embryology into the newly emerging synthesis of evolutionary biology and genetics. These ideas have stood the test of time; they continue to be the best understanding yet generated of how development and evolution are related.

Reclaiming Development

In the decades after Weismann's doctrines decoupled development and evolution, developmental biologists Walter Garstang and Gavin de Beer worked diligently to build a comprehensive theory of biology that would bring about their reunification. When the estrangement between development and evolution finally began to dissolve in the wake of this work, the relations between these phenomena—long assumed to be associated in specific ways—suddenly started to look very different. In fact, reintegrating these phenomena requires us to reconsider our current conceptualizations of evolution itself.

According to the portrayal of evolution that most of us carry around in our heads, evolution results from genetic mutations that directly produce novel, adaptive traits. This belief gives rise to a caricature of evolution as a series of species, one of which is imagined to have evolved into the next:

fish evolved into amphibians when a genetic mutation produced legs (among other things), dinosaurs evolved into birds when a genetic mutation replaced scales with feathers (among other things), and so on. In this portrayal, the *adults* of one species are replaced across generations by the *adults* of a newly evolved species; the image of evolution that most of us maintain rarely, if ever, contains juvenile animals. Noticeably absent from this picture, then, is the *development* that characterizes all living things.

This is not a minor omission, since "evolutionary changes must appear in ontogeny," that is, sometime during development. When you think about it, evolution could hardly work in any other way, given that individuals' traits—the transgenerational alteration of which is what evolution is all about—come to be how they are as a result of *developmental* processes operating during individuals' lifetimes. In fact, evolution is better depicted as a *sequence of ontogenies* than as a sequence of species. "Through the whole course of Evolution," Garstang wrote in 1922, "every adult . . . has been the climax of a . . . life-cycle, which has always intervened between adult and adult. . . . [Evolution] has never been a direct succession of adult forms, but a succession of ontogenies." Specifically, Garstang suggested that evolutionarily novel, adaptive traits arise as a result of *developmental* events that alter an ancestor's characteristics and that are then replicated in the development of the ancestor's offspring.

To illustrate, consider the process by which an imaginary fishlike animal might come to be the evolutionary link between its parents' species and a species of soon-to-evolve amphibianlike animals. Our imaginary fish, like its parents, grandparents, and great-grandparents before him, started off just like each of his siblings: as a fertilized egg that first divided into two cells, then into four cells, and so on, ultimately developing into a small fry, and finally into an adult fish. But as in any population of animals, individuals' varying genes and varying local environments together ensure that individual fish in a school will develop along at least slightly divergent developmental pathways, thereby acquiring at least slightly different traits. Now, imagine that an earthquake throws up an earthen barrier that isolates our school of fish in a new salt lake, one that grows increasingly inhospitable as stagnation reduces the oxygen available in its waters. Under these circumstances, many of the fish will die. While we could conclude that the fish that die did not have the traits needed to survive in their new environment, there is another, richer way in which to view the situation: the fish that die never *developed* the traits needed in their new environment. By the same token, our hero—who finds himself with the idiosyncratic ability to swim to the surface and breathe the air—survives not so much because he can breathe, but because he somehow *developed* the ability to breathe at some point in his lifetime.

This might seem like an incredibly subtle shift in perspective, but it is important because it has the advantage of retaining a role for developmental phenomena in the evolutionary process. If our hero's offspring—and all of their descendants—develop the ability to breathe in the same way that their hero-father did before them, we will ultimately recognize their father as the first of a newly evolved species of air-breathing animals. In such a case, we can interpret what is passed across generations as not a trait per se but as a collection of developmental resources that cause the offspring to follow a particular developmental pathway through life, one that entails the emergence of particular adaptive traits at particular times. According to this portrayal of evolution, novel traits emerge when anomalous *developments* produce animals that differ significantly from their parents; if these animals ultimately reproduce, their offspring can grow up to be like their parents by following the same "abnormal" developmental pathway that their parents followed. In this way, evolution can be seen to involve changes that alter an animal's *development*, changes that are then reproduced in the development of its descendants.

It seems likely that we ordinarily picture evolution in the developmentally empty way we do partly because of how we imagine nature "selecting" individuals for survival. Specifically, many people imagine selection occurring only during life's pivotal moments: when you survive a threat to your life or when you succeed in reproducing, for instance, you have "survived natural selection," you have been "selected." In contrast, a developmental perspective suggests that potentially fertile individuals are being selected by nature in each and every moment that they remain alive. As I have noted, natural selection operates on traits; my current proposal is that over the course of a lifetime, nature effectively selects traits *repeatedly*, in every instant that a trait does not contribute to an animal's immediate demise. As a result, your offspring—by virtue of being developing systems made of genetic and nongenetic factors similar to those that constituted you when you were their age—will develop through a series of moments that are similar enough to the series of moments through which you developed that they will probably survive as you did. Thus, instead of imagining that nature selects well-adapted animals to contribute to the next generation, we would be more inclusive of developmental processes if we acknowledged that what nature *effectively* selects is a sequence of event-moments—the developmental pathway—that *leads to the appearance* of a well-adapted animal. In this way, the development of an animal's descendants will proceed along the lines that so effectively served the ancestor in its lifetime (up to the point at which its descendants were conceived). Thinking of evolution in this way has remarkable consequences, adequate consideration of which could fill another book; here, I will briefly consider only the most important of these consequences.

THE EGG CAME FIRST, NOT THE CHICKEN: CHANGES IN DEVELOPMENTAL RATES CAUSE EVOLUTION

Much of the public believes that some of our traits are caused by genes, that our genes reflect our evolution, and that, therefore, we have these traits for evolutionary reasons. This logic suggests that the development of these traits merely unfolds according to the dictates of evolution, thus recalling Haeckel's view that evolution is "the mechanical cause" of development. In contrast, if we conceptualize evolution as Garstang did—as a process in which evolving descendants traverse different developmental pathways than did their ancestors—our eyes are opened to the possibility that developmental changes might actually *underlie* evolutionary changes; according to this perspective, developmental phenomena might, in fact, cause evolution. Our new perspective literally upends popular thinking.

Gavin de Beer elaborated on Garstang's ideas in 1930, in a short book entitled *Embryology and Evolution*. In this book, and in subsequent editions called *Embryos and Ancestors*, de Beer argued forcefully for the importance in evolution of a process called "heterochrony" (pronounced heter-AH-krony). While this word had been coined by Haeckel decades earlier (and had meant something different to him), de Beer used it to refer to the fact that "the time of appearance of a structure can be altered" from generation to generation. What this meant was that sometimes the development of one of a descendant's traits can be accelerated—causing it to appear sooner—relative to its emergence in an ancestor. Other times, though, the development of one of a descendant's traits can be retarded—causing it to appear later—relative to its emergence in an ancestor. Thus, de Beer argued that in many cases evolution occurs when there is an alteration in the *timing* of the development of ancestral traits. In such cases, an evolutionary modification does not involve the appearance of a completely novel trait but instead "involves no more than a change of timing for developmental stages already present in ancestors."

For example, consider the evolution of the so-called Irish elk, a species of giant deer that, before it went extinct around 11,000 years ago, possessed the largest antlers—ranging up to 12 feet!—of any deer ever known to man. The evolutionary *ancestors* of Irish elks had regular sized, nongigantic antlers when they were adults. But in their descendants, normal-sized antlers were seen earlier in development, that is, in *juvenile* Irish elks. What sort of heterochronic displacement could have led to this evolutionary change?

There are two ways in which traits typical of adult ancestors can come to characterize descendants when those descendants are still immature. First, development of the trait can be *sped up* in succeeding generations so that the trait appears earlier in the development of the descendants than it did in the development of their ancestors. Alternatively, sexual maturity can be *retarded* in succeeding generations, so that development of the trait continues in the descendants beyond the point at which it ceased in the ancestors. Either way, changes in developmental rates produce immature descendants who have traits like those of their adult ancestors. In the case of the Irish elk, it was a retardation of sexual maturity that led to the evolution of its giant antlers; descendants of more moderately antlered deer found themselves to be not yet mature when their antlers had grown to the size of those of their adult ancestors, so they just kept right on growing. By the time they finally reached sexual maturity and stopped growing, their antlers had grown to be colossal.

Heterochronic displacements of the opposite variety—in which traits that were characteristic of *juvenile* ancestors are retained by their descendants into *adulthood*—are also known to be extraordinarily important contributors to evolution. As was the case for displacements of the first variety, there are two ways in which heterochronic displacements of this sort can occur. First, acceleration of sexual maturity in descendants can lead them to stop developing when they still have nonsexual traits that are like those of their immature ancestors. Such a process has contributed to the evolution of several insect species. For example, sexually mature—that is, adult— gall midges look today like their ancestors looked when those ancestors were still immature larvae. Adult descendants can also wind up with traits like their juvenile ancestors when the development of those traits is *retarded* in succeeding generations; in such cases, traits that ancestors "grew out of" by adulthood continue to characterize their adult descendants. Such a process, called neoteny (pronounced nee-AHT-eny), is currently understood to have been pivotal in the evolution of several modern species.

We can see the effects of neoteny in domesticated dogs, which descended from wolves. Adult dogs retain several traits that characterize wolf puppies but not adult wolves. For instance, it turns out that barking, a trait well known to characterize adult dogs, is present in wolf pups who "grow out" of this behavior as they reach maturity. Similarly, the floppy ears that characterize some adult dogs is a neotenous trait that was typical of their wolf ancestors only when those ancestors were pups. But why would the evolution of domesticated dogs be characterized by a *retardation* of development? Actually, this is no great mystery. When early humans selected which wild canines to breed as pets, they chose the ones that most appealed to them. And as the young of many animals are more appealing to people than are mature versions of those animals—people have a harder time resisting a puppy than an adult dog—it is likely that the adult wolves selected for breeding were those that were still somewhat juvenile in appearance even after they had reached sexual maturity. S. Coren, after reporting that research on domestication of the Russian silver fox had supported this hypothesis, concluded "breeding for tameability and friendliness has resulted in dogs that are both mentally and physically more like wolf puppies than wolf adults."

Lest we fall prey to the tendency to look disparagingly at developmental retardation, I should point out that we also see the effects of neoteny in human beings. Gottlieb provided the following illustration:

> An example of neoteny in our own species . . . is our retention into adulthood of the cranial flexure of the fetal period (eyes and nose facing at right angle to spine). This embryonic condition is exhibited by all other mammals, which then deviate [during development] such that the eyes and nose form a more or less continuous line with the angle of the spinal cord in the adult form. The retention of this embryonic feature in [adult] humans is accompanied by others, such as the retention of the fetal skull shape and nonopposable big toe, all of which are necessary to, or are correlated with, upright walking posture.

Stephen Gould is confident that "human beings are 'essentially' neotenous . . . because a *general, temporal retardation of development has clearly characterized human evolution*"; perusal of the available data leaves little doubt as to the astounding similarity that exists between adult humans and juvenile—in many cases, fetal—apes.* Hence, it is now widely accepted among evolutionary biologists that humans evolved from apes when a *slowing*—or in some cases, an arresting—of their development produced the developmental events that generate human characteristics. But why would evolution ever favor retardation in this way? There are a variety of possible reasons, but one that might jump immediately to mind is the fact that such retardation prolongs the period in our lives during which our brains have the remarkably valuable plasticity that characterizes juvenile nervous systems.

Evolution, then, can oftentimes be a *product* of developmental anomalies; in these cases, it involves changes in the *rates* at which development of various traits takes place. Thus, this portrait of the relationship between evolution and development completely inverts how this relationship has traditionally been envisioned. Instead of imagining with Haeckel that evolution causes development, we can now see that, in fact, changes in developmental rates often cause evolution. In Garstang's words, "Ontogeny does not recapitulate Phylogeny: it creates it." This insight led him to some remarkable conclusions, including the inference that "the first Bird was hatched from a Reptile's egg," since bird traits probably first appeared during the abnormal development of a reptile's offspring.* While the significance of this understanding to theoretical biology is unquestionably monumental, I myself am just as impressed with Garstang's ability to finally answer the age-old question: the egg, in fact, must have come first, before the chicken.

Garstang and de Beer had completed their seminal work by the middle of the twentieth century, leaving contemporary biologists with a framework that seemed to rescue development and evolution from the disintegration imposed upon by them decades earlier by Weismann and his followers. Nonetheless, both Garstang and de Beer acquiesced to tenets associated with the modern synthesis, namely that evolution results only from changes in genes; thus, they both "believed that a genetic change or mutation is necessary to bring about the developmental changes that lead to evolution." It is only now, looking at the horizon with a developmental systems perspective, that theoretical biologists are beginning to see the first glimmerings of a truly integrated theory of biology, one that incorporates information about evolution, genetics, *and* development.

SWIMMING MONKEYS AND DEFORMED FLY WINGS: EVOLUTION WITHOUT MUTATION

An important aspect of an integrated theory of biology is that it would permit evolution to result from alterations in *any* of the components of the developmental system, including genetic factors, nongenetic factors, and/ or epigenetic factors that reflect gene-environment interactions. Thus, such a theory would simply require combining Garstang and de Beer's conclusions—evolution can result from developmental anomalies—with the contribution of the developmental systems perspective, namely that development is an epigenetic process involving *co-actions* of genetic and nongenetic factors. The upshot of such a combination would be that *any* adaptive change in development—regardless of what causes the change, as long as it is reliably transmissible across generations—can be evolutionarily significant.

In the conclusion to *Individual Development and Evolution*, Gottlieb argued that "there is so much untapped potential in the existing developmental system (including the genes) that evolution can occur without changing the genetic constitution of a population." In making his case, Gottlieb reasoned that evolution might, in some cases, result from novel behaviors, physiological responses, or anatomical alterations produced in response to changing environmental circumstances.* He wrote:

What needs to happen is the production of animals that live differently from their forebears. Living differently, especially living in a different place, will subject the animals to new stresses, strains, and adaptations that will eventually alter their anatomy and physiology (without necessarily altering the genetic constitution of the changing population). The new situation will call forth previously untapped resources for anatomical and physiological change that are part of each species' already existing developmental adaptability. . . . Evolution involves changes in the [very flexible and highly adaptable] developmental system (of which the genes are an essential part), but not necessarily changes in the genes themselves.

In this passage, Gottlieb is arguing that novel prenatal *or* postnatal developmental circumstances can lead to changes in characteristics, such as anatomical characteristics, that usually seem impervious to environmental influence. But how could this be possible?

Remember, even though experiences cannot ordinarily affect the constitution of the genes, the fact remains that the environment can *effectively* change the genome by influencing which genes are and are not *expressed*, or "turned on." Hence, Gottlieb writes:

The present viewpoint takes advantage of the well-accepted fact that only a very small proportion of an individual's genotype participates in the developmental process. Thus, . . . phenotypic changes can be immediately instigated by a change in an individual's developmental conditions. In our view, a change in the developmental conditions activates heretofore quiescent genes, thus changing the usual developmental process and resulting in an altered . . . phenotype.

Thus, Gottlieb believes that in some cases, novel environmental circumstances can influence genetic *activity* in a way that could lead directly to anatomical or physiological changes. And to the extent that these changes can then be *reliably* passed on to offspring *via any mechanism at all*, they can be subjected to natural selection and thereby contribute to evolution. Ultimately, then, "phenotypic change in evolution need not involve loss [or presumably, gain] of genetic information."

In some ways, Gottlieb's proposal is quite controversial. After all, biological orthodoxy has held for 50 years that evolution results from alterations in DNA itself, and that events occurring during an individual's lifetime cannot possibly affect evolution, since such events cannot breach Weismann's barrier to change an individual's DNA. These ideas follow directly from Weismann's notion of the "continuity of the germ plasm," an idea that was readily accepted by the architects of the modern synthesis and thereby forged into current biological convention. But more than six years after Gottlieb first suggested that nongenetic influences on developmental events could significantly affect evolution, researchers at the University of Chicago reported the following evidence consistent with his proposal.

Working with fruit flies, S. L. Rutherford and S. Lindquist were able to show that when development occurs in conditions that simulate extreme climate changes, a variety of radical anatomical novelties can emerge; in this study, some flies developed "deformed eyes or legs and changes in wing shape." Some of these deformities would likely interfere with a fly's survival in the wild, of course, but others could conceivably be favored by natural selection. In fact, in the laboratory, some of Rutherford and Lindquist's abnormal flies survived to adulthood and were fertile; of most importance, when they were bred with normal flies, some of their offspring developed the same abnormal traits of their monstrous parent. Rutherford and Lindquist inferred that—just as Gottlieb had predicted years earlier— the environmental changes had led to the activation of previously present but previously unexpressed genes. Specifically, they wrote that "widespread variation affecting morphogenic pathways [the developmental pathways by which anatomical traits are produced] exists in nature, but is usually silent. . . . This provides a plausible mechanism for promoting evolutionary change in otherwise entrenched developmental processes." Generalizing from their results and the results of other studies on similar phenomena, Rutherford and Lindquist concluded:

> Experiments disrupting developmental homeostasis by specific mutations [or] by particular [environmental] stresses during precise windows of development . . . have shown that populations contain a surprising amount of unexpressed genetic variation that is capable of affecting certain typically invariant traits. Sometimes very specific [environmental] conditions can uncover this previously silent variation.

The ability of a bird's genes to produce mammalian *teeth* under unusual developmental circumstances—which I will discuss below—drives home the point that major anatomical changes can develop in response to non-genetic signals, even when DNA itself is not altered in any way at all.

When we open our minds to the possibility of evolution originating in altered developmental circumstances, we begin to see how reasonable this idea is, even if the altered circumstances never change DNA, either directly *or* by influencing the ways in which it is expressed. Michel and Moore offered the following example of a way in which novel behaviors that reliably develop in successive generations could influence evolution; this example invokes neither direct anatomical, physiological, or genetic change—or even changes in gene *expression*—but it still underscores the possibility that novel behaviors that develop within an individual's lifetime can nonetheless influence evolution.

Michel and Moore describe troops of Japanese monkeys who live near the beach and who, for the last several years, have regularly received food offerings from human tourists. Because of this arrangement, the monkeys have

> acquired the trait of washing their food, which has led them to spend more time in the sea, which monkeys usually avoid. As a consequence, the numbers and types of marine organisms incorporated into their diet have increased. The animals have also adopted the novel behavioral patterns of swimming and diving.

As a result, should the tourists stop bringing food to the monkeys at some point in the future, the monkeys would not automatically revert to their prehuman niche. Instead, these previously land-dwelling monkeys would have the option of subsisting entirely on seafood or, as Michel and Moore speculate, of "extending their range by swimming to nearby islands." Given the known effects on evolution of the sort of reproductive isolation that such behavior could entail, this example certainly makes it easy to recognize how novel behaviors could drive evolutionary change.

It remains to be seen, of course, if Gottlieb's hypothesis about the possibility of evolution without genetic modification will be supported by empirical data collected in the next few years. Nonetheless, it seems to me to be rational on its surface, and data are beginning to appear that render it increasingly plausible. Moreover, it has the distinction of being a theory of evolution that is not at odds with the known facts of development. Such are the fruits of adopting a developmental systems perspective.

If I have done my job, you are now in possession of three new insights that should alter your thinking about evolution, development, and the relationship between the two. First, genes *and* environments are components of developmental systems that can be reliably reproduced in successive generations. Second, these gene-environment complexes are targets of natural selection (since, according to developmental systems theorists, such complexes effectively *are* incipient traits). Third, *all* traits are acquired. Taken together, these insights allow developmental data to finally be integrated into the 50-year-old modern synthesis of evolutionary biology and genetics. A developmental systems perspective facilitates the elimination of the feature of that synthesis that excluded developmental data, namely the insistence that inheritable traits are determined strictly by genes; in so doing, it leaves us with a more inclusive theory of biology, and at very little cost. Twenty-five years ago, in considering the failure of the modern synthesis to deal adequately with the issue of development, Waddington wrote:

> The entities which undergo evolution are not simply populations of genotypes but are populations of developing systems; that is to say,

organisms one of whose essential features is to undergo development, and moreover, development in which the environment plays a role as well as the genotype. In my opinion the conventional Neo-Darwinist theories . . . are inadequate *both* because they leave out the importance of behaviour in influencing the nature of selective forces, *and* because they attach . . . selective value directly to genes, whereas really they belong primarily to phenotypes [which reflect genetic *and* environmental influences] and only secondarily to genes.

A quarter of a century later, it is certainly time to begin building a theory of evolution that does not suffer from these inadequacies. In the developmental systems perspective, we have a tool that can help us begin.

Part V

IMPLICATIONS

A person without knowledge is surely not good;
he who moves hurriedly blunders.

—Proverbs 19:2

Injustice anywhere is a threat to justice everywhere. We are caught in an
inescapable network of mutuality, tied in a single garment of destiny.
Whatever affects one directly, affects all indirectly.

—Martin Luther King, Jr., in his Letter from a Birmingham Jail

15

WHEN COWS FLY
The Full-Blown Demise of Genetic Determinism

Adopting a developmental systems perspective is likely to change how we think about solving social problems, how we think about assigning credit and blame for people's accomplishments and failures, how we imagine our futures, and—for those of us involved in the biological and social sciences—how we frame our research agendas. It should not be surprising that adopting this perspective would have serious implications; after all, the question of the origin of our traits, both in individual development and in evolution, cuts to the heart of what it is to be both a particular person and a human being in general. I am biased of course—for the last few years I have been living, eating, and breathing the issues addressed in this book—but I honestly cannot imagine a theoretical issue in the sciences that could be of *more* importance to us. Biological development is literally awesome; in a matter of months, developmental processes quite regularly give rise to human brains, the most complex pieces of matter in the known universe. The driving force behind my work on this book has been my desire to share my reverence for such products of nature. Nonetheless, given the potential impact on public policies of our ideas about the origins of our traits, the perspective offered in this book cannot help but have implications that go far beyond theory. The *political* implications of this perspective warrant our close attention as well.*

A NEWLY MINTED "MAP" OF THE HUMAN GENOME, AS DEPENDENT AS EVER

On June 26, 2000, J. Craig Venter, Ph.D., the chief scientific officer and president of a corporation called Celera Genomics, announced at a White House press conference that his company had completed a "rough draft" of a "map" of the human genome. Since the Human Genome Project—the

United States government's attempt to generate such a map—began in 1990 with the goal of discovering the sequence of the human genome by 2005, the scientists were clearly ahead of schedule. At the White House press conference, Tony Blair, the prime minister of Great Britain, said, "For let us be in no doubt about what we are witnessing today. A revolution in medical science whose implications far surpass even the discovery of antibiotics."

Unfortunately, Blair's statement is an example of wild hyperbole. With the discovery of antibiotics, the world was given an immediately useful treatment that really did revolutionize the practice of medicine. In contrast, comprehensive knowledge about the human genome will *not* provide us with any such treatments in the immediate future. As I hope I have now made clear, in the absence of information about the environments in which development takes place, *a perfect and complete map of the human genome will not allow us to make accurate predictions about the traits—or diseases—that a given human being will develop.* Notwithstanding all of the hoopla attending the Human Genome Project, the finished product will leave us much closer to the beginning than to the end of our quest to understand the origins of our biological and psychological traits.

If a complete map of the human genome will not let us identify people who will—or treat people who do—suffer from particular diseases, some might conclude that we are wasting our time mapping the human genome. On this point, I would have to disagree. Since genetic factors contribute to the development of *all* of our traits, a map of the human genome will necessarily be invaluable in our efforts to understand ourselves; without this information, we would be missing a crucial piece of the puzzle. Nonetheless, such a map does not, in and of itself, constitute a completed puzzle, because the appearance of our traits depends on the contexts in which the traits develop. A complete map of the human genome, while certainly a good start, will leave our questions about trait origins unanswered if it is not accompanied by an equally complete understanding of development.

When I have talked over the past few years with my colleagues at conferences, with my students in my classrooms, and with my friends in our homes, they have not been shy about asking me why my understanding of the Human Genome Project's accomplishments is so at odds with what they hear brilliant scientists proclaiming every day in our media. After all, as these scientists have rushed to finish their task, we have been inundated with rosy media claims about how the information being gathered will have immediate practical impacts on how doctors practice medicine. Apparently, Lewontin, too, was being asked a decade ago how he could reconcile his skepticism of this project with the fact that it has attracted enormous amounts of both talent and cash; in 1991 he asked:

Why, then, do so many powerful, famous, successful, and extremely intelligent scientists want to sequence the human genome? The answer is, in part, that they are so completely devoted to the ideology of simple unitary causes that they believe in the efficacy of the research and do not ask themselves more complicated questions. But in part, the answer . . . [has to do with the availability of] economic and status rewards awaiting those who take part in the project. . . . Some farsighted biologists have cautioned against the terrible public disillusionment that will follow [its completion]. . . . The public will discover that despite the inflated claims of molecular biologists, people are still dying of cancer, of heart disease, of stroke, that institutions are still filled with schizophrenics and manic-depressives, that the war against drugs has not been won.

As much as we would all like it to be otherwise, the fact is that *in the absence of an understanding of development,* our new genomic "map" will not help us eliminate these scourges from our lives.

We should not surmise that it is only envious scientists uninvolved with the Human Genome Project who are conveniently conscious of these limitations. In fact, Celera's Venter was interviewed on the ABC News program *Nightline* on the evening of his announcement of Celera's accomplishment; his statements made it quite clear that he is cognizant of the dependent nature of genes. When the reporter interviewing him asked if there were certain genes on which all life depends, Venter replied:

A year or so ago we [at Celera Genomics] tried to take one of the smallest known bacteria and just see if we could get a molecular definition of life. . . . And we knocked out genes one at a time until we got down to about 300 genes and we found that for the simplest cell, there's maybe 300 genes. But there was a problem: there's no unique formula, because it's all context-sensitive. It depends what environment those genes are in. . . . You could only [control the final outcome] . . . if you could control all of the events in the environment.

The implications of Venter's words were apparently lost on the reporter, who later in the interview said, "I have six billion little chemicals in me . . . and some of them are actually producing proteins and making me blink my eyes and talk and stuff." Venter's point was that although DNA plays an important role in protein production, its activity still depends critically on its *context.* And if genetic activity is context-dependent, genes most certainly *cannot independently* cause anatomical or gross physiological traits, let alone behaviors like blinking and talking. We must begin to get beyond the ludicrous idea that our genes can directly control our behaviors, our mental capacities and states, or our aesthetic preferences, because they cannot; this idea is utter pseudo-scientific nonsense.

On the broadest level, a developmental systems perspective frees us from the chains of genetic determinism. This is both good news and bad news, of course. On one hand, the demise of genetic determinism means that no embryo is ever *doomed* by its genetic constitution to a particular future. On the other hand, while some of us hoped that a map of the human genome would end birth defects, disease, and general suffering, such a map will not contribute to this outcome if it is not accompanied by an equally comprehensive understanding of development. Consequently, the explanations and potential solutions offered by developmental systems theorists for society's problems can be counted on to be complex. I will return to these practical points after a first look at the optimistic consequences of genetic determinism's death, a look that will require an understanding of just one more conceptual issue.

THE IMPOSSIBILITY OF INHERENT INFERIORITY

Upon encountering the sorts of data presented in this book, some readers will conclude that developing organisms' traits are more yielding to their environments than they had previously thought. Nonetheless, in my experience, people commonly retain an element of genetic determinism in their thought even in the face of such an insight. Often, this residual genetic determinism appears in the form of a belief that genes determine a *range* of possible developmental outcomes, each of which would result from developing in one of several possible environments. Thus, many of my students arrive in my classes believing that genes bestow a certain *potential* on people. According to this belief, a person with a particular set of genes has the potential to grow to an adult height between, say, 64 and 68 inches; what *actual* height the person attains within this range, though, is determined by nongenetic, developmental factors. Given this arrangement, a person with these genes could never grow to a height beyond this range, no matter what the person's developmental environment. This idea is denoted by the common textbook phrase "range of reaction." And while such a notion does retain roles for both genetic and nongenetic factors in trait development—and so is well founded compared to full-blown genetic determinism—the data of developmental biology have, in fact, rendered it effectively invalid. Instead, for all practical purposes, genes do not even determine a *range* of possible developmental outcomes.

The idea that genes determine a range of reaction was traced by S. A. Platt and C. A. Sanislow to I. I. Gottesman, who wrote, "a genotype determines an indefinite but circumscribed assortment of phenotypes, each of which corresponds to one of the possible environments to which the genotype may be exposed." In other words, if you let genetically identical clones develop in different environments, you will get a range of different developmental outcomes. Thus, while Gottesman understood that genes do not *determine* traits independently of nongenetic factors (as evidenced by his use of the word "indefinite"), he clearly thought that they *do* determine a range of possible outcomes (as evidenced by his use of the word "circumscribed"). Astonishingly, Gottesman came to this conclusion even though the reference he cited to support his claims—a 1955 book titled *Evolution, Genetics, and Man* by Theodosius Dobzhansky—*specifically rejected the possibility that the "range-of-reaction" idea has any practical value at all.*

Eight years before Gottesman loosed the range-of-reaction idea on the world, Dobzhansky—one of the architects of the modern synthesis—wrote, "we can never be sure that any of these traits have reached the maximal development possible with a given genotype. The performance of a genotype cannot be tested in all possible environments, because the latter are infinitely variable." Dobzhansky understood that to know the full range of traits that a given genotype would be capable of producing, we would need to let an organism with that genotype develop *in every possible environment* and observe the traits that develop. So, while a range of reaction might be discoverable *in theory* if a collection of clones could be allowed to develop in all possible environments, such a range cannot *in fact* ever be specified, since environments are infinitely variable. As Dobzhansky so insightfully pointed out in 1955, "The existing variety of environments is immense, and new environments are constantly produced. Invention of a new drug, a new diet, a new type of housing, a new educational system, a new political regime introduces new environments."

Thus, we can never conclude, for any practical purposes, that a genotype *circumscribes* a range of possible developmental outcomes. Instead, all we can ever know for sure is what the most *common* developmental outcome is for a given set of genes. Consequently, developmental biologists have rejected the idea that genes specify a *range* of possible developmental outcomes—a range of reaction—in favor of the idea that genes are associated with *typical* developmental outcomes in specific environments—a *norm* of reaction. The distinction between a range of reaction and a norm of reaction might appear at first to be rather subtle, but, on further reflection, its significance will become clear.

The first important thing to note about norms of reaction is that each genotype has its own. For example, 9,000 clones of Marilyn Monroe, each raised in a different environment, have the potential to grow to differing heights as a function of variation in environmental factors that influence height (such as diet). Similarly, 9,000 clones of Humphrey Bogart, each raised in one of the *same* 9,000 different environments in which we raised our Monroe clones, will also grow to a variety of heights. But the fact that the norm of reaction associated with Bogart's genes need not be the same as that associated with Monroe's genes means that the *differences* between their heights will not be the same across the different environments. Instead, while many adult Bogarts will be taller than the adult Monroe who developed in the corresponding environment, there are liable to be some environments in which the Monroe clone will grow to be taller than her Bogart counterpart. Lewontin, S. Rose, and L. J. Kamin made this point with the less fanciful example of corn clones:

> One genotype may grow better than a second at a low temperature, but more poorly at a high temperature. . . . [M]odern corn hybrids are superior to those of fifty years ago when tested at high planting densities in somewhat poorer environments, while the older hybrids are superior at low planting densities and in enriched conditions. Plant breeding has then not selected for "better" hybrids. . . . Thus genotype and environment interact in a way that makes the organism unpredictable from a knowledge of some average of effects of genotype or environment taken separately.

What this means is that *we can never assume that changing the conditions in which two people develop will affect both people in similar ways.* As a direct consequence of this fact, there are no such things as genes that are invariably associated with certain developmental outcomes; it will always be possible that in *some* environment, a zygote produced by tall parents will develop into a shorter adult than will a zygote produced by short parents. I'll return to the practical implications of this observation in a moment.

The second important thing to note about norms of reaction is more obvious: they preclude the possibility of stating outright that there are certain things of which particular genes are not capable. This sounds like a radical statement, because we are all quite confident that human genes are not capable of producing wings. But look again: How can we *know* that this is an absolute limitation on human genes if no one has ever tried to find an environment in which these genes might contribute to the growth of wings? As ridiculous as this sounds, the fact is that we cannot know such a thing in advance. This conclusion is supported by a truly bizarre study that shows how inappropriate it is for us to assume things about the limitations of genes before actually studying their reactions in *all* possible developmental circumstances.

Two decades ago, E. J. Kollar and C. Fisher conducted an absolutely incredible experiment that demonstrated just how unpredictable genes' reactions can be. Following in the footsteps of Hans Spemann—who, you will recall, conducted transplantation studies that led to the creation of two-headed newts, thereby demonstrating that one type of tissue can *induce* the differentiation of another type—these researchers conducted transplantation experiments with a group of five-day-old chick embryos. Ordinarily, a chick's beak develops out of a particular layer—let's call it layer 1—of embryonic tissue. Specifically, the layer underlying layer 1— call it layer 2—induces layer 1 to differentiate into beak tissue. This is the same sort of process underlying the differentiation of *teeth* in mammal's mouths: one layer of tissue induces the layer overlying it to begin differentiating into teeth. As shocking as it might sound, Kollar and Fisher first removed the layer 2 chick cells from the developing chick embryos and then replaced them with transplanted layer 2 *mouse* cells taken from 17-day-old mouse embryos. Unbelievably, in the context of the layer 2 mouse cells, the layer 1 chick cells actually differentiated into complete, well-formed mammalian *teeth*! Consider for a moment how truly remarkable this is: the cells that became mammalian teeth were *normal chick cells*, complete with normal chick DNA, and *only* chick DNA. What this means is that in the right context, chick genes—which *never* generate teeth under ordinary circumstances—are capable of collaborating with nongenetic factors in their local environments to co-construct mammalian teeth with "perfectly formed crowns . . . [and] with root development in proper relation to the crown."

How in the world can this be possible? Kollar and Fisher noted that the evolutionary ancestors of birds—widely thought to be dinosaurs—had teeth, the point being that the genetic factors that contributed to the construction of teeth in those ancestors might possibly have been retained (in a silenced form) as dinosaurs evolved into birds. To paraphrase their rather technical language, Kollar and Fisher noted that the ability of layer 1 chick cells to differentiate into teeth suggests that the normal toothlessness of modern birds is "not a consequence of a change in the genetic coding [during evolution]. Rather, an upset of a developmental sequence . . . must have blocked the initiation of tooth development." They concluded by referring to the fact that genes can lurk unexpressed in our chromosomes for very long periods of time, ready to spring into action should factors in their environment suddenly turn them "on."

Kollar and Fisher's results, then, underscore how wrong it is to assume there are limitations on a given set of genes before studying their reactions to all possible developmental circumstances. And as Dobzhansky realized nearly 50 years ago, since it is not possible to study all possible developmental circumstances, we can never confidently assert *anything at all* about genetic limitations. This understanding led Michel and Moore to point out that, as far as we know, "it may be within the possibility of the human and cow genome to support the development of wings, just as it is within the possibility of chickens to grow teeth." Thus, the notion that our genes specify a restricted range of possible developmental outcomes can be of no practical use whatsoever.*

Given these observations, the seemingly subtle difference between the range-of-reaction idea and the norm-of-reaction idea turns out to have significant implications for social policy. Of most importance, it renders ideas about genetic superiority and inferiority nonsensical. And I'm not speaking here only of a fascist-flavored *general* genetic superiority or inferiority but rather also of a genetic superiority or inferiority in *any one* of the specific traits that characterize a person. A century of scientific research on trait development in clones* has delivered a clear verdict: genes that contribute to the construction of an inferior trait in one environment can contribute to the construction of a superior trait in a different environment. Thus, it is not possible to conclude that particular genes will invariably produce a taller person than will other genes, because *it will always be possible that some environment exists in which genes that contribute to superior height in most environments contribute to inferior height in this one.*

Of course, the implications of this insight fly in the face of all of our deeply ingrained "common sense" understandings of trait origins. Consider the following implication, for example: given what we now know, there can be nothing *inherent* in Michael Jordan's *genes* that makes him a better basketball player than I am. This follows from the fact that there might be an environment that could have constructed a better basketball player in collaboration with my genes than in collaboration with Mr. Jordan's.* Thus, as crazy as this assertion sounds, the data of developmental biology affirm its truth. It simply is not the case that we can assume that a given environmental manipulation will have equivalent effects on everyone exposed to it, so a manipulation that improves my basketball playing would *not necessarily* just turn Mr. Jordan into even more of a superstar. By the same token, increasing regimentation in a classroom *might* improve or harm every child's developmental outcome, but it could also help some and harm others. The only way to know what impact a particular manipulation will have on a variety of different people is to try the manipulation and see what happens.

The raw data of developmental biology are clear: it is not possible to identify *any* individual as "genetically inferior" or "genetically superior" to any other individual, even in a specific domain. Moreover, the conclusion that a set of genes can never accurately be labeled "inferior" or "superior" does *not* reflect technological immaturity of the life sciences. Instead, when the human genome has finally been completely mapped—we currently have only a "rough draft"—it will *still* be impossible to conclude that particular genes will *always* produce traits of a particular quality. Consequently, the origins of the limitations that currently frustrate disadvantaged people will probably not be found by searching among their genes. Rather, the origins of those limitations that exist probably can be found by improving our knowledge of developmental processes.

Given our current ignorance of developmental processes, and given the fact that it is simply impossible to identify people who are *genetically* unable to benefit from access to social resources like quality education and nutrition, it seems incumbent upon democratic societies to distribute these resources equitably. The fact that genetic information alone will *never* be able to specify which people would benefit *most* (or least) from access to these resources merely serves to reinforce this exigency.

16

BUDGING THE BELL CURVE
Implications for Education

The publication in 1994 of *The Bell Curve* by Richard Herrnstein and Charles Murray sparked a firestorm of controversy that resulted from these authors' provocative conclusions that intelligence is largely heritable and that there are undeniable differences in the average IQ scores of black and white people. While Herrnstein and Murray never actually stated outright that the inferior IQ scores of black people are caused by inferior genes for intelligence, many scholars who read their long book took this to be the implication of their arguments. The social policy recommendations that invariably flow from such arguments are obvious: the distribution of social resources via government welfare or education programs such as Head Start is inefficient, as some people are less able to benefit from such expenditures simply by virtue of their genetic endowment.

Where I stand on this contentious issue will not surprise anyone: since I believe that even biological traits in simple animals are influenced by both genetic and nongenetic factors, I am confident that human intelligence is similarly influenced by both types of factors. But even Herrnstein and Murray would have conceded this point. They claimed, not that environmental factors do not contribute to intelligence, but that studies of the *heritability* of intelligence show that variation in IQ scores is *better* "accounted for" by variation in genetic rather than environmental factors. As we now know, however, estimates of the heritability of intelligence tell us *nothing at all* about the relative contributions of nature and nurture to the development of this trait. In fact, we currently know virtually nothing about the ways in which genetic and nongenetic factors co-act to influence the development of intelligence; this is why I have not yet addressed this issue in this book. Nonetheless, the importance of this debate—coupled with my own strong sense that a developmental perspective can shed some light on what would otherwise likely remain a fiery battle—impels my following comments.

The Black, White, and Gray of IQ

There continues to be debate among reasonable people about whether or not IQ tests actually measure intelligence; they might or might not. Regardless of the final disposition of this debate, however, there can be little doubt that, *on average*, Americans of African ancestry have traditionally scored lower on these tests than have Americans of north European ancestry (but see research by D. Tate and G. Gibson for contradictory evidence). And because a person's IQ score is a relatively good predictor of many other factors that are important to people (for example, success in school, occupational achievement, socioeconomic status), it is reasonable to attempt to understand the origin of this "trait" *even if it in fact has nothing at all to do with the trait we intuitively know as intelligence.*

Is it possible that measured differences in IQ among black and white people reflect genetic differences between these groups? Absolutely. Is it possible that this difference reflects differences in the environments in which members of these groups develop? Absolutely. Is it possible that one of these groups possesses a genetic superiority in the domain of intelligence? As should now be clear, this is a possibility unworthy of consideration, because such a superiority could not possibly be ascertained. For all practical purposes, it is *impossible* to accurately conclude that a person—let alone a group of people!—has genes that render him genetically inferior to another person, either in general or in particular domains like intelligence. Nonetheless, it does remain possible that particular sets of genes are at a disadvantage *in particular environments.*

As it happens, there *are* well-documented differences in the average developmental environments of people of African and European ancestry. Halford H. Fairchild, for example, points out that "blacks, on the average, suffer from less access to health care, obtain less prenatal care, and live in more impoverished and stressful residential areas than do whites." Similarly, E. Lieberman has reported that black babies are almost three times more likely than white babies to be born with low birth weights. It is widely agreed that these differences account for some of the measured difference in these groups' IQ scores. But what accounts for the rest of the difference? Behavior genetics studies have consistently estimated the heritability of IQ *among white people* to be about 70 percent, meaning that genetic differences among them "account" for most of the variation in their IQs. Thus, it is commonly assumed that *genetic* differences between blacks and whites, too, are responsible for some of the difference in their average IQ scores.

Before considering this possibility, it is important to recall that the high heritability of IQ among whites does not necessarily mean that IQ is similarly heritable either among blacks or in a mixed population of blacks and whites. Heritability is a statistic that is different in different populations, a fact that led Block to point out, "There is no reason to expect the heritability of IQ in India to be close to the heritability of IQ in Korea." Therefore, it would be inappropriate to assume that IQ is similarly heritable in white, black, and mixed black-white populations.

Nonetheless, *even if we accept the contention* that IQ is similarly heritable in these populations and that, therefore, genetic differences do "account" for some of the difference in blacks' and whites' IQ scores, this still does not mean that black people's genes are responsible in *any* direct way for their lower IQ scores. A large heritability estimate for IQ *could* reflect a direct genetic contribution to IQ, but it could just as easily reflect the fact that black children, on average, are likely to have inferior educational experiences relative to white children. It works this way because if dark skin is associated with poor education, and poor education is associated with low IQ scores, then a gene that contributes *directly* to the development of dark skin leads—however *in*directly—to poor performance on IQ tests. And according to the correlational logic underlying heritability studies, the educational link in this chain can be ignored, since, in fact, in this scenario, genes for dark skin allow reasonably good prediction of IQ scores. Those who are concerned only with heritability would simply ignore the fact that the implicated genes have no *direct* impact on IQ at all.

Finally, it is important to remember that even if all people were similarly advantaged and there were no such things as inferior educational experiences, a perfect, 100 percent heritability estimate for IQ would still not mean that genetic factors are more important contributors to IQ than nongenetic factors. This is true, you will recall, because heritability estimates tell us only about *differences* in people's traits, not about the *causes* of the traits themselves. The significance of this point is that *environmental manipulations can have overwhelmingly powerful effects on IQ even if genetic factors alone accounted for* all *variation in human IQ scores*. The source of an IQ *difference* that arises when different genotypes develop in different environments—which is what heritability estimates measure—is *simply irrelevant* to questions about the malleability of IQ in the face of environmental manipulations. Thus, regardless of the source of the IQ difference between blacks and whites, a high heritability *cannot* be taken to mean that manipulations of the educational environment are wasteful. Changes in environmental factors can lead to changes in IQ, *regardless* of IQ's heritability.*

Even behavior geneticists who study the heritability of IQ concede that environmental manipulations can significantly influence IQ scores. For example, it is common knowledge among those familiar with adoption studies that when children from poor socioeconomic backgrounds are adopted into more well-to-do families, their IQ scores are ultimately much higher, on average, than their biological parents' scores. In addition, in *The Bell Curve*, Herrnstein and Murray discuss a phenomenon they called the "Flynn effect," after the author of a 1987 study that found that IQ scores in several of the world's countries had risen about three points, on average, in every decade since World War II. Thus, since then, the average "IQ in many countries has gone up 15 points, about the same as the gap separating blacks and whites in the USA." While we do not yet know enough about environmental influences on IQ to fully explain this phenomenon, the "Flynn effect" clearly reflects an *environmental* influence on IQ, since the human gene pool has not changed dramatically in the past 50 years. Obviously, the high heritability of IQ among white people in no way interferes with the malleability of their IQs in the face of environmental changes.

Thus, manipulations of nongenetic factors can produce changes in IQ. But might the high heritability of IQ be taken to mean that this trait is *relatively difficult* to affect with environmental manipulations, even if it *can* be done? As we saw in an earlier chapter, this is not the case; heritability estimates do *not* tell us about the ease with which a trait can be influenced by environmental manipulations.* Instead, *if you understand how a trait develops*, influencing highly heritable traits with environmental manipulations is *just as easy* as influencing traits that are not particularly heritable. (By the same token, it is equally difficult to affect heritable and nonheritable traits in predictable ways if you do not understand how they develop.) Let me repeat this crucial point: *The ease with which a trait can be affected by an environmental manipulation is not a function of the trait's heritability, but it is a function of our knowledge about how the genetic and nongenetic factors that co-construct this trait interact during its development.* Thus, just as heritability estimates cannot help us *explain* the development of a trait, they are of no use to us if our goal is to *influence* its development. Both affecting and understanding the IQ disparities that characterize black and white populations require knowledge about how IQ *develops*. In contrast, heritability estimates offer us no guidance as we struggle with what to do about these disparities.

WHAT GOOD ARE HERITABILITY ESTIMATES, AND WHY ARE THEY STILL CALCULATED?

Given what heritability estimates do *not* tell us, can we take knowledge of the heritability of IQ (or of any other trait) to have *any* practical value at all for anyone besides rhetoricians? Definitely. Heritability estimates continue to be useful in helping to predict the outcomes of selective breeding programs (because we can hold environments constant in these situations). As such, these estimates continue to be of practical use to both farmers and

Nazis (or eugenicists, more broadly). In addition, these estimates might be useful to a third group of people, namely those who wish to make decisions about their *own* procreation based on their own or their partner's traits; heritability estimates might be of some very limited value to parents concerned about passing their own afflictions on to their children.* Outside of these three groups of people, I am not aware of anyone who would find any practical value at all in heritability estimates; frankly, the rest of us can—and should—ignore these numbers completely and still rest assured that we are not overlooking any useful information.

Unfortunately, heritability estimates are not merely useless for most purposes; they also can actively impede scientists' inclination to search for the *causes* of traits among *developmental* processes. As Daniel S. Lehrman noted as long ago as 1953, concluding that a given trait is "heritable" (or, for that matter, "innate," "genetic," or "instinctive") has the consequential effect of inhibiting further investigation into the *development* of the trait. In Lehrman's words, such conclusions "inevitably tend to short-circuit the scientist's investigation of . . . relationships which underlie the development of 'instinctive' behavior." This is because labeling a trait "instinctive" or "inherited" has the immediate effect of seeming to provide *the* answer to the question of the trait's origin. And while this is, in fact, not an answer at all—but instead, an admission that we do not understand how or why the trait developed—it often strikes us as a reasonably satisfactory explanation, allowing us to shrug our shoulders and move on to different scientific problems. It is, perhaps, in this consequence of heritability calculations that we find the most damage being done to the search for genuine understanding of trait origins.

If heritability estimates cannot guide us as we struggle with how to address black-white IQ disparities, how should we try to improve the IQ scores of children whose lives we believe could be improved by such interventions? Unfortunately, as I have noted, we know very little about the developmental events that give rise to the skills that contribute to successful IQ test performance. This ignorance argues only for redoubled efforts to understand the *development* of these skills, not for acquiescence to a current state of affairs. In the meantime, if we *know* that high IQ scores are associated with improved qualities of life, if we *know* that environmental factors affect IQ scores, and if we are not about to alter anyone's DNA in the interest of improving his or her IQ score,* it makes sense to do whatever is in our power to raise our children—*all* of our children—in environments that we *think might be* conducive to intellectual growth. Until such time as our ignorance of intellectual development recedes in the face of our efforts to understand it, it seems only fair that we encourage our leaders to distribute society's resources equitably, so that the developmental environments of different embryos, fetuses, babies, children, and adolescents are as equivalently "enriched" as possible.

I feel obliged to emphasize at this juncture that, as far as I know, the policy implications associated with the developmental systems perspective *follow* from focusing on the fact that traits develop; I do not believe this scientific worldview emerged from a particular political orientation. Nonetheless, the political implications of this view do seem to be a factor in the scientific (and broader) community's resistance to it.* P. J. Bowler's thoughts regarding the modern synthesis are relevant here:

> There may be genuine scientific reasons for suspecting that the synthetic theory [that is, the modern synthesis] does not offer a complete explanation of evolution, but most biologists feel that the alternatives will only command their respect if they can be dissociated from the anti-Darwinians' often one-sided pronouncements on moral issues.

While Bowler's observations are not about developmental systems theories—these theories are not anti-Darwinian—they do highlight how preexisting political positions can interfere with the acceptance of new scientific ideas. They might also help us understand why heritability estimates are still calculated in an age when we know that such estimates tell us nothing at all about either traits' origins or their malleability.

I believe there are genuine scientific reasons for adopting a developmental systems perspective, reasons that are completely independent of "moral issues." Nonetheless, one's perspective on trait origins cannot help but give rise to a political position. If genes predetermine traits' appearances, there is little reason to try to influence trait development with political policies, since such attempts would be doomed to fail. But if traits develop as a result of gene-environment co-actions, we are encouraged to try to influence the emergence of those traits that our society wants to abet (for example, intelligent behavior) or control (such as violent behavior). The fact that a scientific theory might give rise to particular political positions should not prevent this theory from gaining the respect of the scientific community, if there are genuine scientific reasons that this theory should command that respect. Any insistence that new scientific perspectives warrant respect only if they are dissociated from the political, moral, or otherwise practical implications of those perspectives is misguided. We must not let our moral preconceptions blind us to the potential value of scientific theories that call these preconceptions into question.

17

FINDING THE COST OF FREEDOM
Complexity and Responsibility

Adopting a developmental systems perspective leads to the heartening conclusion that people are never genetically doomed. Nonetheless, this perspective also carries with it some mildly discouraging principles. In contrast to the attractive simplicity of the synthetic theory's approach, the developmental systems approach is discouragingly complex, and its adherents remain skeptical of claims that complicated medical or social problems can be treated effectively with simple solutions. If these adherents are correct, we can expect that vanquishing our ignorance of developmental processes will take a lot of hard work. Moreover, as developmentalists see it, we are all responsible for *all* of our behaviors; there are no such things as behaviors that are controlled by our genes. But given how people are drawn to simple answers that hold out the possibility that we are not responsible for what we do, the naïve notion that genes cause traits is now as firmly entrenched in the public's consciousness as ever.

Consider the recent trend for constituents of our society to "geneticize" behaviors that they deem abnormal or otherwise undesirable. For example, it is quite common these days to hear assertions about the genetic bases of behavioral characteristics like homosexuality, violence, drug abuse, or learning disabilities. Often coupled with these assertions is the implied suggestion that these characteristics, by virtue of being "genetic," are more like uncontrollable diseases than like lifestyle choices or character weaknesses that people *should* control. Of course, these characteristics *must* reflect, somehow, the genetic constitutions of the individuals who develop them, but they *cannot* be thought of simply as "genetic," since the study of development has shown unequivocally that genes cannot directly control behaviors.

There is a long history associated with the argument that differences between people are rooted in their biological characteristics. In considering why genetic explanations are often more attractive than social explanations for behaviors that some segments of society find troubling, Alper and

Natowicz concluded that "as societies . . . became stratified, biological explanations of differences among ethnic groups and between the sexes were used to account for and justify the inferior social status of women and minority ethnic groups." More specifically, they write:

> Slavery in the United States was justified by theories of the innate intellectual and moral inferiority of African Americans. The Nazis carried these theories to their ultimate conclusion; millions of Jews, Romanies (gypsies), Slavs, and homosexuals, who were determined by the German scientists of the day to be genetically degenerate, were killed to protect the "racial purity" of the German people.

Thus, biological explanations for alleged "problematic" traits often reflect "the attitudes and values associated with the more politically powerful groups in society." More recently, however, some of the less politically powerful groups in society—homosexuals, for instance—have thought it advantageous to use such explanations themselves to understand the appearance of their own controversial traits. As a result, it is not uncommon these days to hear people disavowing control of some of their own behaviors, maintaining instead that these behaviors are controlled by their genes (or are otherwise biologically "fixed").

It is obvious why politically powerful people would classify as "biological" those traits that they consider inferior: it provides a way for them to rationalize less-than-equal treatment of those possessing these traits. While fascist "solutions" are an extreme example of such biased treatment, less extreme examples can still be found in modern America. For instance, the argument that intelligence is largely genetic implies that it is reasonable to distribute society's educational resources inequitably, since those with "inferior genes for intelligence" are less able to benefit from access to these resources anyway. But what can explain the recent attraction of less powerful minorities to "biological" explanations for their own traits? Alper and Natowicz offer a few suggestions, including the possibility that such explanations are relatively simple and therefore easily understood, that they imply that we might discover "cures" for the traits (an attractive prospect to some with these traits), or that they might encourage society to see these behaviors as being out of an individual's control and more resistant to change than behaviors that seem more socially determined. "It is hoped, perhaps naïvely, that this third reason, in turn, will have the positive social consequence of removing blame and discrimination against individuals having the difference in question."

A developmental systems perspective, in contrast, will not allow us to disavow control of our behaviors simply by labeling these behaviors "genetic." This is because developmentalists understand that each of our behaviors is a complex product of a physical brain, and that the structures and functions of brains reflect both the environments in which they develop as well as the characteristics of the genes they contain. And if *all* behaviors are influenced by both genetic and nongenetic factors, it makes little sense to consider some behaviors "genetic" in a way that others are not.

Behavior can never be understood to reflect only *either* psychological *or* biological factors; this follows from the fact that brains and behaviors mutually influence one another. Behaviors emerge from a brain, but a brain's structures and functions are themselves influenced by the experiences that result from one's activities and circumstances. Thus, brains and the behaviors they generate always reflect both biological *and* psychological factors. Arguments over whether or not behaviors are "biological" (usually implying that these behaviors are somehow immutable) or "psychological" (usually implying that these behaviors reflect choice or explicit learning on the part of the actor), can invariably be seen to be bogus arguments.

Consider, for example, the need for speed. Racing skiers probably have brains that differ from those housed in the rest of us; we don't *all* get the same rush from skiing at extremely high velocities, and those of us who do get such a rush don't all react to it in the same positive way. But whenever a developmentalist hears that someone has an unusual brain, she immediately asks the developmental question: How did this person's brain *come to be* as it is now? If the brains of fast skiers differ from those in the skulls of the rest of us, it *cannot* be for reasons that are completely experience-independent, since a brain *always* reflects both its genes *and* its experiential history, broadly construed.

The same argument holds for other traits. Drug abuse *must* reflect both biological and psychological factors that together have produced a brain that craves certain drugs. After all, experience with drugs *always* precedes addiction to drugs, and an addiction certainly influences subsequent experiences. Ultimately, there is little reason to think that drug abuse is any more of a genetically produced "disease" than is fast skiing; in both cases, a brain forged by gene-environment interactions motivates its body to seek out particular kinds of stimulation. Similarly, homosexuality is unlikely to be any more or less "psychological" or learned than heterosexuality. Given that sexual orientation is produced by a brain, it cannot help but reflect a complex interaction of genetic, hormonal, cultural, and other factors, since all of these factors influence brain development. Thus, our abilities to control our own drug use, sexuality, or need for speed are not a simple function of how "genetically determined" these impulses are. Instead, because each of these impulses reflects a brain state that has developed through gene-environment co-actions, the role genetic factors played in their development is not relevant to questions about how difficult the impulses are to control.

This is not to say that all of our behavioral characteristics are under our conscious control; behaviors obviously vary in this regard. Nonetheless, it is a mistake to think that behaviors that are difficult to control are any *more* influenced by genetic factors than behaviors that are relatively easy to control. After all, I find it easy to control the movement of my eyeballs and difficult to control my self-deprecating thoughts, but I am as sure that the latter reflects my childhood experiences (among other things) as I am that the former is influenced by characteristics of my genome (among other things). Whether or not a behavior can be easily controlled is not a simple function of the contribution genetic factors make to the appearance of that behavior.

Faced with this understanding, developmental systems theorists specifically reject the validity of any "my genes made me do it" explanations for behavior. It remains possible, of course, that people who have certain traits—for example, a strong motivation to consume alcohol—are genetically different from those of us who do not have these traits. Alcoholics might, in fact, share certain genes that the rest of us do not. But genes alone cannot be held *responsible* for behaviors, because behaviors are never determined by genes alone; environmental factors are always involved in behavioral development.* In the end, behaviors known to be associated with the presence of particular genes are just as amenable (or resistant) to self-control (or to modification by other people) as are behaviors not known to be associated with particular genes.

The attraction people have for simplistic biological explanations of certain traits, then, ultimately appears to be ill founded. First, scrutiny of biology always shows it to be anything but simple, the wishful thinking of those who resort to biological explanations notwithstanding. Second, while various behaviors might ultimately be rendered controllable by "cures" derived from biological research, there is little reason to think that such "cures" will be more easily found—or in the end, any more effective—than "cures" based more explicitly on traditional psychological principles. After all, while medical interventions certainly affect our psychological processes, psychological interventions, likewise, affect our brain structures and chemistries. Generating a comprehensive understanding of the processes by which troublesome behaviors *develop* should lead to the identification of a variety of possible "treatments" for these behaviors, treatments that need not be rigidly classified as "biological" or "psychological." Finally, the belief that biological explanations for certain traits can be effectively used to either justify certain political programs or to encourage tolerance of behavioral diversity is almost certainly misguided.

Despite these arguments, crude biological explanations for traits are as widely accepted among the public today as they have ever been, propped up by journalistic simplifications and wishful thinking. If such explanations were merely incorrect, perhaps our society could tolerate their widespread appeal. Unfortunately, however, these explanations are not only incorrect, but also carry with them imprudent prescriptions for dealing with social maladies.

THE ILL-FATED (AND PROBABLY ILL-CONCEIVED) FEDERAL VIOLENCE INITIATIVE

To see the consequences of taking the simplistic path favored by many (but eschewed by developmentalists), consider the Federal Violence Initiative proposed by President George H. Bush's Alcohol, Drug Abuse, and Mental Health Administration in the early 1990s. The goal of the initiative was to treat urban violence as a public health problem; that is, the goal was to find ways to identify and provide treatment for potentially violent criminals *before they had committed any serious violent crimes*. Armed with a nondevelopmental approach to the origins of violent behavior, the initiative had at its disposal a potentially simple solution to this exceedingly complex problem: given what is (and is not) known about biological correlates of violent behavior—in particular, that violent behavior in primates is correlated with low levels in the brain of the neurotransmitter serotonin—one of the schemes available to the initiative would have entailed screening urban youths for low brain serotonin levels and providing serotonin-boosting medications to those young men deemed "in need" of treatment. The initiative came under fire in 1992—and was subsequently abandoned—shortly after its point man, Frederick Goodwin, likened impoverished urban youths to monkeys by drawing an analogy between America's ghettos and the jungle. In addition, Goodwin had expressed his belief that biological factors "inclining human beings toward violence . . . might be [used as] 'biological markers' of violent disposition." Part of the public outcry over the initiative also reflected people's concern that the initiative might endorse the medical treatment of otherwise healthy, innocent people.

Given how I think about the origins of traits, *I* was certainly pleased to see the initiative derailed. Nonetheless, I remain concerned, as do many Americans, about violence in our country; I continue to believe that this violence warrants efforts to curtail it. It would be convenient, of course, if there *were* a simple solution to this problem, but research into the *developmental* origins of violent behavior has—not unexpectedly—revealed a complex network of causes underlying its emergence. Specifically:

> We know that habitual aggression is a multiply-determined behavior,
> with a range of individual, situational, contextual, and historical factors
> contributing to [it]. . . . For instance, we know that . . . low birth weight,
> exposure to environmental toxins, and neuropsychological trauma increase
> an individual's risk of behaving violently. It is also the case that individual

social skill deficits, particularly poor problem solving and anger management skills, are associated with aggressive and violent behavior. We know that aggressive behavior is more likely the result of repeated observation of aggressive models and reinforcement for aggression. . . . We know that violence occurs most often in communities and cultures that tend to condone fighting and endorse aggressive norms. We know that aggressive and antisocial behavior is more likely to be present if parents do not monitor their children's activities, and less likely to be present if families provide a warm and caring environment. We also know that teenagers who are involved in conventional activities and provided with adequate job opportunities are less likely to engage in violent behavior.

Thus, despite the complexity of this problem, the circumstances that contribute to violent behavior *can* be identified. The problem of violence is not intractable.*

Obviously, research on the developmental origins of violence has revealed a process of inordinate complexity. But if our society's values lead us to demand solutions to this problem, developmental analysis will be able to provide them. Developmental psychologists are currently exploring ways to prevent violence and other forms of delinquency in adolescents; the more successful programs being studied involve providing mothers with prenatal care and children with intensive preschool enrichment experiences. Other developmentally sophisticated researchers are looking into the possibility that aggressive behavior can be curtailed by teaching children how to non-violently resolve their conflicts with their peers. Once we have a better idea of how to prevent the development of violent behavior, we will be able to provide parents and teachers with information and materials they can use at their discretion to encourage their children to develop into less violent adolescents. Such an approach would preclude the possibility of governments drugging children simply because their brains are characterized by neuro-transmitter levels that, while common among violent people, may *not* play a causal role in the development of violent behavior. Never forget: correlation tells us nothing at all about causation.

As is the case with most of the stubborn problems that plague our communities, learning how to reduce violent behavior will not be easy.* Still, I prefer the trepidation that comes from facing a lot of work to the demoralization that comes from facing an unsolvable problem. Fortunately, there is no merit to the idea that our biology has us straightjacketed, poised to be sent ballistically into an unyielding, predetermined future. We are each responsible for our own behaviors, so it behooves us to learn as much as we can about how to control them. It is time to throw our full support behind studies of the *development* of behaviors and other traits that are of concern to our societies.

18

MEDICAL MARVELS
Implications for Genetic and Environmental Therapies

In *Darwin's Dangerous Idea*, philosopher Daniel Dennett tells of a discussion he had with Francis Crick, one of the scientists who discovered DNA's structure. During their conversation, Crick declared that psychologists who make extraordinarily simplified computer models of human brains are doing "terrible science." Because population geneticists (who, among other things, study correlations between specific traits and specific genes) also work with extraordinarily simplified models—in this case, of DNA—Dennett decided to see if Crick would similarly denounce population genetics.* Dennett writes:

> The derogatory term for some of their [population geneticists'] models is "bean-bag genetics," for they pretend that genes for this and that are like so many color-coded beads on a string. What they call a gene . . . bears only a passing resemblance to the intricate machinery of the codon sequences on DNA molecules. . . . Is their research good science? Crick replied that he had himself thought about the comparison, and had to say that population genetics wasn't science either!

While Dennett believes that population genetics *is* a perfectly good science, Crick apparently thinks otherwise. Presumably, this is because he is well aware that there are not really such things as "genes for this and that," and that models built upon such ideas are liable to lead us farther away from genuine understanding.

It turns out that the deeper you read into the professional medical literature, the less likely you are to find overstated claims about the primacy of genes in the production of our traits. Instead, the popularity of the idea that genes cause traits originates mostly in the provinces of professional journalism and secondary education. Perhaps this reflects a failure on the part of scientists to convey the dependent nature of genes to interested journalists and teachers, or a failure on the part of scientists, journalists, and teachers to appreciate the extent to which their audiences can understand it. Regardless, as one reads the work of scientists directly involved in the study of genes and their effects, it becomes obvious that the dependent nature of genes is widely appreciated in this community.

THE CASE OF HEREDITARY HEMOCHROMATOSIS

A concrete example of this appreciation can be found in the story of hereditary hemochromatosis (pronounced heem-oh-krome-a-TOH-sis). This is a relatively common disorder* in which dietary iron is absorbed to an excessive degree by the stomach and intestines, allowing abnormal amounts of iron to accumulate in the body. When the body's ability to store unneeded iron has been exhausted, the excess iron is deposited in a variety of organs, ultimately leading to liver damage, heart failure, diabetes, generalized fatigue, arthritis, abdominal pain, impotence, and/or death, among other things. In 1996, "the gene for" hemochromatosis (or, more accurately, the gene that, when mutated in particular ways, contributes to this disorder) was reported to have been discovered. This discovery heralded the possibility of identifying those healthy people likely to suffer the symptoms of this disorder later in their lives; this is significant insofar as early treatment of the disorder can effectively leave sufferers with a normal life expectancy.* Thus, the pattern of events in this case was much like that to which we have all become accustomed: high-tech research in a molecular biology lab reveals "the gene for" a disorder, the discovery is announced amid great fanfare, and optimism runs high that effective treatments are imminent.

Then, in the summer of 1998, the Centers for Disease Control and Prevention (CDC) and the National Human Genome Research Institute (NHGRI) convened a panel of experts to discuss the practical implications of the discovery of "the gene for" hemochromatosis. In particular, the panel considered the potential value of screening the general population for mutations of this gene, in the hope that early identification of those with these mutations might improve their chances of remaining symptom-free. In the end, however, the panel concluded that "genetic testing is not recommended at this time in population-based screening for hereditary hemochromatosis." The question, of course, is why *not* screen for the presence of these mutations if they really are *the* mutations that *cause* hemochromatosis?

It turns out—and this will not surprise those of you who have slogged your way through this book—that "simple detection of the mutation does not predict the most likely clinical course [of the disorder]." Instead, among homozygotes—that is, people who carry *two* copies of the recessive mutation, one on a chromosome originating with their mother and the other on a chromosome originating with their father—symptoms range all the way from "none that are detectable" to "severe organ damage from iron overload." Thus, people who have *two* of these abnormal genes—and so who should definitely show signs of the disorder according to the simplistic genetics we are all taught in high school—still do not necessarily develop the symptoms of hemochromatosis. As a result, the panel wisely concluded that "uncertainties . . . about the percentage of homozygotes that will develop [symptoms of the] disease and the effect of genetic or environmental modifiers on [the presence of these symptoms], indicate the need for caution in the use of genetic testing." This is just another example of the now-worn truism at the center of this book: genes do not cause traits independently of the environments in which development occurs.

Given this truism, the best way to screen for abnormal traits is to turn our attention to the traits themselves and away from exclusive focus on the genes that are touted as causing these traits. An example of a test developed with this approach is the standard test used to screen newborn babies for PKU. Although PKU is widely considered by nondevelopmentalists to be a genetic disease (my protestations notwithstanding), when we screen newborns for PKU, we do *not* look for babies who have the genes associated with PKU. Why is that? Because the genetic mutations associated with PKU do not *invariably* produce this disorder. Instead, genetic testing would miss some babies who would ultimately develop PKU, while simultaneously misidentifying some babies as likely PKU sufferers even when these babies just process phenylalanine in a trivially abnormal way. Thus, when we screen newborns for PKU, we look not for genes, but for babies who *in fact* have abnormal amounts of phenylalanine in their bloodstreams.

Faced with a possible genetic test for hemochromatosis, the CDC's panel of experts concluded that genetic testing for this disorder would be inferior to direct tests of blood iron levels, just as genetic testing for PKU is inferior to direct tests of blood phenylalanine levels. Specifically, the panel wrote, "there may always be an inherent uncertainty in a hemochromatosis diagnosis made on the basis of genotype rather than on evidence of increased iron stores." They concluded the section of their report on the implications of testing for "the hemochromatosis gene" with the following observations:

> As we acquire more knowledge about the molecular basis of genetic disease, it becomes increasingly clear that variable expressivity (i.e., modification of a genetic trait by other genes or the environment) is the rule rather than the exception.

and,

> It can be argued that a phenotypic screening test is always preferable to a [genetic] test, because phenotype is the clinical concern. In this view, the availability of a phenotypic test should lead to increased caution in the use of a DNA-based test.

Of most importance, it should be noted that these last statements are absolutely *general*, and appropriately so. After all, in contrast to most traits (be they normal or abnormal), disorders like PKU and hemochromatosis seem to involve the activity of very few genes. If, as the panel's report acknowledges, genes do not single-handedly cause even *relatively* uncomplicated "genetic" disorders, it is surely a mistake to believe that accurate genetic tests for more complex disorders (or for normal traits) are around the technological corner.

Ultimately, studies of the *development* of both normal and abnormal traits will yield better prescriptions for society's ills than will nondevelopmental studies of these same traits. Remarkably, this statement is true *whether we are talking about traits within the traditional realms of physicians (for example, disease) or psychologists (for example, intelligence or violence)*. While many people understand that the latter sorts of traits reflect our experiences, we have all been slow to accept the idea that the former sorts of traits—some diseases in particular—are just as "open" to environmental influence. But those scientists who actually study genes know better: even "genetic" diseases are caused by genetic factors that co-act with nongenetic factors. Given this basic truth, the cause of any trait—normal, abnormal, psychological, or biological—can be understood only via study of its development.

GENE THERAPIES, ENVIRONMENTAL THERAPIES

Given that genes cannot single-handedly cause abnormal characteristics, it might be less than surprising that researchers have been largely unsuccessful in their attempts to develop gene therapies. But success can always be just around the next corner. On April 29, 2000, the front pages of both the *New York Times* and the *Los Angeles Times* reported what might turn out to be the first ever successful use of a gene therapy to treat a disease: French scientists alleviated the symptoms of severe combined immunodeficiency disease X1 in three afflicted infants.* The successful gene therapy involved removing stem cells (which I will discuss shortly) from the infants' bone marrow. The researchers took these cells, installed in them a sequence of DNA bases that code for a particular protein, and then returned them to the infants' bones. One year later, all three infants appeared to have normally functioning immune systems. Jennifer Puck of the NHGRI has said that it is "too early to say if they have been completely cured, but for now, they seem to be doing just as well as if they had been cured."

Such results certainly support the belief that a decade of gene therapy failures does not warrant abandoning research on genetic contributions to diseases. After all, it is no wonder that getting gene therapies to work has been extremely difficult; as Katherine High of Children's Hospital of Philadelphia points out: "It has proven to be a much more complex undertaking than was initially imagined." Even so, gene therapy research is almost certain to yield marvelous medical breakthroughs; since we all develop in environments that are remarkably similar in regards to certain crucial factors (for instance, the presence in utero of oxygen, gravity, glucose, and so on), genetic differences can be expected to be highly correlated with differences in developed traits. Thus, how certain genes interact with "normal" environments to produce certain disorders is almost guaranteed to be useful information. The fact that research on genetic contributions to disease has thus far produced only one successful therapy should not deter us from continuing to study genes.

But given the epigenetic nature of development and the potentially low cost of environmental therapies relative to gene therapies, we would be wise to augment the time and money we spend seeking *basic* understandings of how nongenetic factors contribute to development. Such understandings are as likely as are understandings about genes to lead to practical applications and the development of new therapies. Many such therapies will prove to be both less expensive and at least as successful as therapies based on our recently won understandings of genes. We must not be seduced by the lure of novel high technologies *if* older approaches coupled with novel understandings can fulfill the same functions more efficiently. While it is likely that the future incidence of heart disease, for example, will be reducible via genetic interventions, the fact remains that nongenetic "interventions"—altering diet, stress, tobacco consumption, or activity level, for example—can reduce the incidence of heart disease *now*, and at a fraction of the cost.

One of the general principles that developmentalists have already wrested from nature's grip is that interventions early in development can be more effective than later interventions in influencing long-term developmental outcomes. The reason for this can be understood metaphorically by picturing a Waddington-style "epigenetic landscape" containing two relatively deep valleys. Once a ball rolling down this landscape has entered one of the two valleys, its final destination is relatively predictable; it would be difficult (though not impossible) for the ball to jump the watershed separating the two valleys and come to rest in the valley it did not initially enter. Switching metaphors, if an array of sequentially falling dominoes reaches a fork in the road that leads to the falling of the left, and not the right, domino, the chances of dominoes further down the right hand pathway ever falling become increasingly small. Developmentalists capture the spirit of this phenomenon when they say that the *developmental history* of an organism constrains its current traits *and* the future course of its development.* This is not to say that later developments are *un*important—events occurring in your thirties certainly affect characteristics you have as a 60-year-old—but, in general, events occurring earlier in life play an important role in specifying the rough contours of development.

Because traits are always constrained by developmental history, they can be influenced most dramatically by genetic or environmental manipulations that are carried out early in development. This realization has facilitated the design of therapeutic interventions that can be implemented early in life and that can therefore have far-reaching consequences. For instance, being deprived as a baby of patterned visual stimulation—as happens when cataracts cloud an infant's vision—leads to abnormal development of the brain's visual areas. To date, such abnormal development has proven irreversible when normal vision is not restored early in childhood; thus, timely detection of visual abnormalities is essential. However, as traditional opthalmological methods rely on self-reports about what patients can see, for years no one could conceive of a test of a *preverbal* infant's visual acuity. Without such a test, of course, there was often no way to identify babies in need of therapeutic intervention before such time as they could *tell* us about their blurry vision. These children typically passed their infancies in a visually obscured world, deprived of experiences that all normal children share; often, these abnormal experiences left them with permanently abnormal brains. Among the contributions of my infant perception colleagues of which I am most proud is the invention of techniques to assess the visual acuity of preverbal infants. Without detailing how this is done, suffice it to say that these techniques now allow for the early identification (and, in some cases, treatment) of abnormal visual development, thereby saving many children from blindness. In general, the earlier in development we attempt our manipulations, the more likely they will radically alter development's course. This state of affairs argues strongly that we not hesitate to spend time, money, and energy on studies of early development; utilizing resources in this way is liable to generate valuable dividends for our society in the end.

THE PROMISE OF EMBRYONIC STEM CELL RESEARCH

The study of development can also lead to unexpected insights and therapies. For example, there have been unanticipated but important practical consequences of the discovery that undifferentiated cells in very young embryos—which do not yet have much constraining developmental history behind them—have the potential to become *any* type of cell contained within an adult body.* So-called embryonic stem (ES) cells, which possess this remarkable property, can be obtained from embryos that are still just clumps of undifferentiated cells. Three years ago, a team of researchers in Wisconsin became the first scientists to successfully culture a collection of ES cells in a laboratory dish. The significance of this achievement rests on the hope that these cells, when cultivated in quantity, could be implanted into people suffering from cell degeneration in specific organs. Once implanted, it is believed that ES cells might be able to differentiate into needed cell types, thereby halting the progression of the degenerative disease. This is an idea that is still in a theoretical (and in some cases, experimental) phase, so it remains to be seen if it will have the practical therapeutic applications scientists hope it will. So far, though, the data from studies of undifferentiated cells have been encouraging.

For example, John Gearhart has reported that "ES-derived cells have been successfully transplanted into fetal and adult mice," whereupon they took on forms and functions appropriate to the location in which they were implanted. He continues:

> On the basis of the use and study of mouse ES cells, the research and clinical potential for human ES cells is enormous. . . . It is exciting to speculate on how human ES lines could be used in tissue transplantation therapies. Obvious clinical targets would include neurodegenerative disorders, diabetes, spinal cord injury.

D. Solter and Gearhart speculate further, writing, "One can even imagine using existing organs as scaffolding, replacing the original cells with those derived from ES cells."

A related line of research has already had some controversial success in the treatment of Parkinson's disease. This disorder—a degenerative neurological condition characterized by tremors, muscular rigidity, slow movements, poor balance, and difficulty initiating or ceasing movements—has been known since the 1950s to be associated with the death of neurons in a brain area called the substantia nigra. In 1990, a group of scientists collaborating in Sweden, England, and Switzerland attempted to exploit the fleeting, unconstrained potential of relatively undifferentiated fetal cells by implanting them into the substantia nigra of a Parkinson's disease sufferer. Remarkably, these researchers found that the implanted cells restored neurotransmitter synthesis in the patient's brain, and that "this neurochemical change was accompanied by a therapeutically significant reduction in the patient's severe rigidity."

Despite their exciting success, this research group's transplantation treatment remains controversial for several reasons. First, its effectiveness has not proven to be universally replicable. Second, some control patients, who have not been subjected to brain surgery, have nonetheless experienced as much improvement as that experienced by transplant recipients (such so-called placebo effects are common, but they are currently inexplicable). Third, the long-term effects of implantation are currently unknown. Fourth, it might be that similar improvements can be obtained using less risky surgical—or in some cases, even nonsurgical—manipulations. Finally, fetal brain tissue transplantation research remains controversial because of the *source* of the cells used in these studies. This last concern warrants an additional moment's attention.

Unlike ES cells, which are obtained from the formless cell mass that constitutes an embryo before it implants itself in the uterus, the cells used by the Swedish-English-Swiss team were obtained from the brains of two-month-old human fetuses whose mothers had voluntarily ended their pregnancies. Thus, these cells were not stem cells with limitless developmental potential; instead, they were *relatively* undifferentiated neuroblasts, already committed to developing as *some* type of brain cell.* Transplantation of cells obtained from aborted fetuses' brains has struck some observers as ethically questionable, to say the least. "Specifically," K. A. Crutcher writes:

> The need for viable (living, survivable) fetal brain tissue requires that the fetal brain be alive, at least at the tissue level. . . . The possibility that the acquisition of fetal brain tissue involves vivisection . . . raises the question of whether anesthesia might be necessary to avoid additional, unnecessary suffering on the part of the fetus. . . . Furthermore, the issue of consent, which . . . involves informed consent of both donor and recipient, has not been adequately addressed in the case of fetal tissue donation. The ethical dilemma arises from the fact that the mother, who consents to the death of her offspring, cannot be assumed to represent the interests of that offspring. As a result, current guidelines for procuring fetal tissue do not follow the long-established tradition of including safeguards for the protection of the donor.

Another concern is the possibility that individuals, if they *know* that fetal brain cell implantation might help those suffering from neurodegenerative disorders, might actually conceive a fetus specifically in order to donate its brain tissue for use in a transplantation protocol. It is not difficult to imagine a 20-year-old woman choosing this path in an effort to save her 45-year-old father from the ravages of Parkinson's disease.

Clearly, both the ethical and scientific problems associated with this type of research need to be satisfactorily addressed before undifferentiated cell transplantation techniques can be used appropriately in clinical settings. Nonetheless, these techniques certainly hold as much potential as do gene therapies for the treatment of human disorders. It is reasonable to expect continued study of early developmental processes to yield additional therapeutic advances. Ultimately, studies of how genetic and nongenetic factors *co-act* to build our traits will reveal developmental secrets that will improve the quality of our lives.

But it is not only scientists exploring trait origins who need to understand the importance of studying development. The same ideas also demand the attention of nonscientists, as we are all responsible for electing those officials who will decide how the government will distribute our hard-earned tax dollars to the country's scientific laboratories. After all, when the Human Genome Project has finally completed a map that specifies exactly which 97 percent of our DNA does nothing at all, and which 3 percent can code for which proteins, American citizens will have spent about $3 billion on the endeavor. Given the relative success of therapies based on studies of the *development* of our biological and psychological traits, I believe such research should be as heavily funded—by both public and private sources—as is current research on the genes associated with these traits. A developmental systems perspective leaves us encouraged to continue seeking both environmental and gene therapies for behavioral and medical disorders.

19

OWNING CLONING
Implications for New Reproductive Technologies

In the spring of 1998, researchers at Scotland's Roslin Institute announced the arrival of Bonnie, a newborn baby lamb. Like any other lamb born on a Scottish farm, Bonnie was conceived by her mother and father and subsequently carried to term and delivered by her mother. As usual, then, Bonnie's birth represented the beginning of a new life cycle; to date, her development has been the picture of normality. There is only one thing that distinguishes this story from the story that unfolds on thousands of farms across the world each day: Bonnie's natural mother was Dolly, the first animal ever cloned from an adult mammal.

We have already learned quite a lot from Bonnie's mom and all of the other clones who have followed her. Our newspapers and magazines do their best to keep us up to speed with this astoundingly fast-moving new reproductive technology. In addition, there have been several good books published recently that consider its ethics, meaning, and potential. In concluding this book, I will focus only on the issue with which I began: the idea that a clone is identical to the donor animal that provided its genes. It is in this widespread belief that we can see how strongly we still hold ideas rooted in genetic determinism.

The fact that genetic maps do not specify developmental outcomes implies that cloning adult animals is *not* tantamount to creating living copies of those animals, popular perception notwithstanding. Instead, when Ian Wilmut cloned an adult ewe at the end of 1995, what he really did was create a copy of a *zygote*, displaced in time. This zygote then developed in natural ways, ultimately *becoming similar* to the adult ewe that was cloned. Think of it like this: if someone were to attempt to clone me, they would proceed by removing one of the cells constituting my body. Then they would starve the cell of nutrients; this step is an environmental manipulation that brings the cells' chromosomes into a state like that which characterizes the chromosomes of a newly formed zygote. Finally, they would fuse it with an unfertilized egg cell that had previously had its entire nucleus removed, chromosomes and all. The product of this procedure would be a zygote that is a copy of *the zygote that I was* 40 years ago when my parents first conceived me. The problem with the idea that cloning produces copies of adult animals can now be seen to be the same problem with which we have been wrestling throughout this book: traits reflect the *environments* in which they develop as much as the genes that contribute to their construction. Thus, to wind up with a copy of me, the copy of the zygote that *became* me would have to *develop in the same environmental circumstances in which I developed*.

In some ways, this is possible, since many qualities of the uterus in which I developed are regularly reproduced in women today; if my zygotic clone were artificially implanted into such an environment—that is, into a surrogate mother—it is likely that in nine months a baby would be born with many of the traits that characterized me as a baby. However, even this newborn would not be identical to the newborn that I was, because no environment is *exactly* the same as the one in which I developed 40 years ago. My mom smoked tobacco and drank alcohol while she was pregnant with me—the potential consequences of these behaviors were not yet well understood in 1960—and it is unlikely that the surrogate mom carrying my zygotic clone today would do the same. No doubt this surrogate would also have a different diet than the one my mother had (I don't think you could find sushi or pad thai in North Carolina in 1960). Furthermore, there is good reason to think that the air the surrogate would breathe and the water she would drink would differ from those taken in by my mother. And it is highly unlikely that the levels of sex or stress hormones in my clone's "mother" would be identical to those you could have found 40 years ago in my mother; people just vary too much in this regard. And the list goes on. In many important ways, then, it is not possible to provide my cloned zygote with the *same* environment in which I developed as a fetus; that environment simply no longer exists.

But do these differences really matter? In some cases, perhaps not so much. The environmental factors that contribute to the appearance of highly canalized traits are liable to be so ubiquitous in human uteruses that these traits—ten fingers, ten toes, two eyes—can be counted on to characterize my newborn clone. But in *many* other cases, these variations probably *will* make an important difference. The data presented earlier reveal how influential circulating sex and stress hormones can be on genetic activity and development, and how significant dietary factors can be in the development of body mass. And while I was unable to find space in this book to discuss how developing fetuses are affected by maternal exposure to pollution or drugs like tobacco and alcohol, the deleterious effects of such exposure are now well known. My clone might be delivered from its "mother" resembling the baby that came to be me, but under no circumstances could it be considered to be an identical copy of that baby.

Obviously, this is just the beginning of a long story, as I certainly continued to develop after I was born. And just as my fetal environment could be reproduced today only in part, my postnatal environment, too, is largely gone; Cleveland of the mid- and late-1960s is history. Some aspects of my postnatal environment could still exist and characterize my clone's postnatal environment, of course; these factors might interact with my clone's genes to produce some traits like those I have developed in my natural lifetime. There can be little doubt that my 13-year-old clone would be similar in some ways to how I was when I was 13. But because there would be *many* differences in our developmental environments—each of which could affect the quality of our traits—there can be little doubt, too, that my 13-year-old clone would be a unique individual, easily distinguishable from the 13-year-old that I once was. As the developmental pathway of my clone diverges from the pathway that I followed as a 13-year-old, the differences between us could only be expected to increase, ensuring my clone's uniqueness as an adult. Given that so-called identical twins are *never* identical in nature, we can be confident that a pair of clones developing alongside each other in real time would likewise grow up to be distinguishable individuals; it can hardly be possible, then, for a clone developing 40 years later than its "DNA-donor/parent" to grow into a *copy* of that "parent."

On February 11, 2000, David Baron reported on National Public Radio that scientists, who by that point had cloned dozens of sheep, cattle, and mice, have in some cases been surprised by the variability that characterizes genetically identical clones. Given the need to dispel the widespread belief that clones are identical copies, I quote at length from Baron:

> Take George, Charlie, and Albert, for instance. The two-year-old Holstein calves . . . were [cloned] from the DNA of a single animal. The calves do look similar. They're all the same size . . . each is mostly black with a white triangular splotch on his forehead. But where Albert and George have a white spot above their shoulders, Charlie has a stripe stretching all the way down his right side. White patterns on the animals' legs are different, too.
>
> University of Massachusetts biologist Jim Robl created the clones. "We've had a lot of people look at them and it's always a question: these aren't identicals! In fact, I had a person in the, ah, beef industry, that was trying to convince me that these could not be identicals because they don't have the exact same coat-color pattern." Robl says he expected

subtle differences in the animals' coloration, because the patterning is due to more than just genetics, in the same way identical twins have different patterns of freckles and different fingerprints. But Robl says his cloned calves also have distinct personalities.* "When they were younger, and there were lots of opportunities for people to come out and take pictures of them, Albert was always much easier to work with than George and Charlie were, because Albert was always more laid-back than the other two."

At the University of Hawaii, researchers have found that mice cloned from a single individual don't always grow the same. Some of the mice become normal sized adults while others become obese, despite being fed the same diet.

And at the Roslin Institute in Scotland, the creators of Dolly also created four Dorset ram clones. . . . The institute's Ian Wilmut says the rams are noticeably different in size and temperament. "There are big differences in their behavior, with some being much more assertive than others."

While the prevalence of such variations in clones seems to have surprised some scientists, a developmental systems perspective leads us to *expect* such variations; in fact, it would be *uniformity* of appearance or personality that would be surprising. Baron let Keith Campbell of Wilmut's team explain this to his listeners:

> Development is controlled by . . . instructions in the genome, but it's also controlled by instructions that are received in the egg and by organelles that are only inherited from the mother in the egg. So if you're using . . . eggs from different individuals to make clones, then there are most likely to be differences in some of the phenotypes that are observed.

"Add to that," Baron said, "each cloned embryo is implanted in a different surrogate mother, which provides a different environment for the growing animal. . . . Biologists say clones will never be truly identical because there's so much more to an individual than one's genetic makeup."

Scientists who produce clones have not yet learned to control the natural developmental processes that build mature animals. Instead, these scientist have learned only to create—from "spare parts," not from "scratch"—*zygotes* that, *when placed in specific environments, develop in the same sorts of currently unexplained ways as do naturally conceived zygotes.* The fact that we do not yet have the know-how to control the production of plants or animals in the way that we currently control the production of objects will, of course, be a relief to some people and a bane to others. But given what we *do* know about the critical roles of nongenetic factors in trait development, we can *expect* every organism to be unique; clones will *not* be like Xerox copies of one another, even if they share identical genes. Why? Because *genes do not determine phenotypes*.

To understand the origin of our traits, we will have to continue to explore the human genome, of course; since our genes influence *all* of our characteristics, understanding how they work will be a necessary part of any comprehensive portrait of development. Nonetheless, a developmental systems perspective urges us to devote equal time to studying gene-environment *co-actions*; understanding the human genome is just one step in a long journey to understand how our traits originate in developmental processes. Only with a thoroughgoing understanding of these processes will we succeed in our efforts to understand why we develop the characteristics we do.

EPILOGUE
In the Beginning

The viewpoint offered in this book frees both individuals and society from the chains of genetic determinism; when the epigenetic origin of traits is more widely appreciated, no longer will people have to bear the thought that their destiny and the destiny of their children is preordained in their genetic endowment and forever beyond their dominion. Understanding how various factors contribute to trait development fosters a sense of empowered hope that is dimmed by the ignorant acceptance of genetic determinism.

With this sense of optimism, though, comes an equally important responsibility. Since "how we are"—human nature—can be understood from a developmental systems perspective to result from interactions between factors *all of which we can conceivably manipulate*, we are ultimately responsible for our own nature. In this sense, developmental systems theory effectively challenges the very existence of an eternal "human nature," since we cannot *be* any one way at all if all we need to do to change our nature is to change our developmental experiences. As Lewontin has pointed out, Simone de Beauvoir had it right when she observed that "a human being is 'l'être dont l'être est de n'être pas,' the being whose essence is in not having an essence." Because of this state of affairs, the implications of the developmental systems perspective are extraordinarily wide-ranging, encompassing how we think about education, personal responsibility, the distribution of society's resources, medical therapies, and reproductive technologies, among many other things.

Despite the importance of the developmental systems perspective and the fact that it has been at hand for over a century—since Driesch demonstrated that early embryonic cells are equipotential, and so *must* be affected by extracellular factors as they develop—it continues to be underappreciated, by both the public and some scientists. As recently as 1994, a prominent geneticist could still be quoted—in the premier scientific journal of our time, *Science*—making developmentally nonsensical statements such as "raise a Newfoundland with border collies, and it will never have [the ability to maintain eye contact with people]; these things are pure genetics, and they're absolutely uncontroversial." If scientists who should know better can speak out in such ignorance, can it be any wonder that journalists who translate scientists' statements for public consumption often pass on similarly fallacious ideas?

Consider the events that unfolded in the mid-1960s when researchers first suggested that abnormal men with one X *and* two Y chromosomes (XYY men) were more aggressive than normal men: the association was reported extensively but uncritically in the media. Subsequently, as C. C. Mann has pointed out, "other researchers attacked the data; a 1993 report from the National Academy of Sciences dismissed the link as unproven." Nonetheless, the tendency of journalists to excitedly report not-yet-proven associations between traits and genes has continued unabated into more recent times. Mann writes:

> The pattern continued into the 1980s with scientific reports linking reading disability to genes on chromosome 15, schizophrenia to chromosome 5, psychosis to chromosome 11, and manic depression to chromosome 11 and the X chromosome. All were announced with great fanfare; all were greeted unskeptically in the popular press; all are now in disrepute.

In considering how to grapple with this situation, Scott Gilbert recently addressed the question of who needs to be made aware of the ideas born of the developmental systems perspective. He wrote:

> Two churches on which to post these theses would be those housing behavioral geneticists and journalists. These two groups interact synergistically, each positively feeding back on the other. The newspapers trumpet that behavioral geneticists have identified such things as the "gay gene," the "alcoholism gene," and the "depression gene." Afterwards, when other laboratories cannot replicate these findings and the scientific reports are quietly withdrawn, the public is not told about the withdrawal. Our culture subscribes to a molecular phrenology [a reference to the nineteenth-century pseudo-science of inferring from the shape of a person's skull things about her mental faculties] and genetic causation sells. Why do people "want" to hear that genes are responsible for behaviors? . . . On one level, genetic determinism becomes a useful way both to avoid responsibility and to define certain groups or individuals as being naturally bad (or good). On a deeper level, genetic causation appears so compelling because that many people believe that genes are the essence of our identity.

Exploring related ideas, D. Nelkin and M. S. Lindee point out that genes have "become a cultural icon, a symbol, almost a magical force" and that in popular culture, they function "in many respects, as a secular equivalent of the Christian soul." It is not difficult to see how genes have taken on this function; studying biology at this level cannot help but inspire a sense of awe for the miracles of nature. But, ultimately, this sort of "genetic essentialism" "reduces the self to a molecular entity, equating human beings, in all their social, historical, and moral complexity, with their genes." Given the data presented in this book, reducing ourselves in this way clearly reflects erroneous understandings of the workings of biological systems, as awesome as they are. People are quite a bit more than their genes.

The antidote to genetic essentialism, of course, is a clear understanding that in all thriving biological systems, a gene *depends* for its function on the activity of other components of the system, components that exist in the gene's—or broader—environment. And how can we get our hands on this antidote? By viewing with appropriate skepticism reports of gene-trait correlations and, instead, turning our attention to the origins of traits in *development*. The emergence of such a change in society's focus would be revolutionary.

ACKNOWLEDGMENTS

Few of the ideas presented in this book are my original contributions to an understanding of trait development. Instead, most of these ideas have grown from the fertile minds of persevering scientists who are not satisfied to shrug, complacently mystified, in the face of the miracle of life. These are people driven by their gifts for asking just the right probing questions, by their curiosity, by their love for nature and its mechanisms, and, in many cases, by their desire to improve the human condition through increased understanding.

My interest in development stems from my exposure in college to research on infants. As a senior psychology major at Tufts University, I was lucky enough to meet Michael J. Weiss, a graduate student who at the time was working toward his Ph.D. with Dr. Philip R. Zelazo at the Tufts–New England Medical Center. Having been well advised by Weiss that psychology graduate school admissions committees look most kindly upon those applicants who have had some exposure to research, I was primed to accept his offer when he told me he was in need of a research assistant. As it happened, Weiss and Zelazo were studying memory in human newborns, and Weiss was in need of an extra pair of hands and eyes to help him collect data for his dissertation. I cannot have known how my life would change from the decision to volunteer in Zelazo and Weiss's infant laboratories. I had never even *seen* a newborn before, let alone considered the possibility that scientists might be able to figure out some of what goes on in their heads. But one trip with Weiss to the newborn nursery at St. Margaret's Hospital in Boston was enough to convince me that focused study of development could ultimately answer one of the most important questions facing humankind: How do we get to be how we are?

I continued to study development as a graduate student at Harvard University, where I received wonderful training at the hands of Drs. Jerome Kagan and J. Steven Reznick. Through graduate school, I became increasingly interested in the biological underpinnings of psychological phenomena, and so began to study the relationships that exist between brain developments and the development of both behavioral and other psychological characteristics. Ultimately, it was this direction of study that led me to the ways of thinking that I have presented in this book.

Before taking my current position at Pitzer College and the Claremont Graduate University, I spent one year doing a postdoctoral fellowship in Manhattan at the City University of New York (CUNY) Graduate Center. One of the most stimulating parts of this program turned out to be the weekly meeting of CUNY developmentalists, at which we would discuss the recently published work of an agreed-upon developmental psychologist. In 1988, Dr. Katherine Nelson was running the psychology program at the Graduate Center; as best I can recall, it was Dr. Nelson who suggested we discuss "The Persistence of Dichotomies in the Study of Behavioral Development," by Timothy D. Johnston (1987). At the time, I found this article to be fascinating, but, honestly, I can't say that I was able to fully understand how revolutionary—and how useful—Johnston's rejection of dichotomies truly was. As I described in the introduction, it was another few years before my colleagues Robin P. Cooper and Robert Lickliter shared with me the developmental insights that led me to ultimately adopt the nondichotomous views I have presented here.

I would not have been able to write this book without the help and support of colleagues, friends, and family, many of whom read and offered me comments on earlier drafts of this work. Among those who shared their time and minds with me as I worked to get these ideas on paper were Gilbert Gottlieb, Mita Banerjee, Robert Lickliter, Jerome Kagan, Stephen Rudicel, Ken Stahl, Paul Bloomfield, Robin P. Cooper, Glendon Good, David Barash, the students in my Seminar on Individual Development and Evolution (Leah Levine, Consuelo Bingham-Mira, David Byer, Ashley Denault, Rachel Fay, Matt Fehrs, Lorena Gonzalez, Greg Preston, Megan Pulham, Kori Shaiman, Doug Wein, Stephanie Silberstein, and Laura Cocas), André Ariew, Chris Moore, Mishtu Banerjee, an anonymous reviewer, and my family—my father Dr. Irwin B. Moore, in particular. I am also grateful to the colleagues at Pitzer and throughout the Claremont Colleges who provided the supportive environment in which I was first able to start accommodating to these ideas, including Alan Jones, Halford Fairchild, Richard Tsujimoto, Norma Rodriguez, Leah Light, Jeff Lewis, Karen Kossuth, Jim Bogen, Judy Grabiner, Mary Gauvain, and the Claremont Colleges Developmental Reading Group (Patricia Smiley, R. Lee Munroe, Marguerite Malakoff, Claire Kopp, Sheila Walker, and Michelle Wierson). In addition, I have been inspired by my teachers—Josh Bacon, Michael J. Weiss, Philip R. Zelazo, Dante Cicchetti, J. Steven Reznick, and Katherine Nelson—and by many other theorists on whose work I have drawn heavily in my writing, including Gerald Turkewitz, Esther Thelen, Linda B. Smith, Scott Gilbert, Steven Jay Gould, Richard Lewontin, George Michel, Celia Moore, Susan Oyama, Mark H. Johnson, Elizabeth A. Bates, and Timothy D. Johnston. I would also like to thank those friends who either graciously allowed me to share their anecdotes with the world or allowed me to share mine with them, when needed: James Taylor; Dona Bailey; John Kolligian; Lynn Whitney; Callie Rodriguez-Morken; Dr. Claire Tilem and her twins, Steven and Daniel; Elisa Mendel and Hayley Falk; Robert and Michael Jones; Sam Lieberman; Nick and Beverly Muto; Kim and Robert Gatof and their son, Jake; and Pamela and Eric Finkelman and their son, Sam. Finally, I would like to thank Julia Hough, Elizabeth Knoll, and Susan Arellano from Cambridge, Harvard, and Yale University Presses, respectively, for encouraging me in my early efforts to find a publisher for this work. In this regard, I am particularly indebted to my colleague Barry Sanders for helping me find an agent, my agent Miriam Altshuler for helping me find a publisher, and my editor at W. H. Freeman, Erika Goldman, for believing in the value of this project.

Notes

Introduction

5. A school environment that is "good enough": Scarr, 1992.
5. *In April 2000, a jury in a Connecticut superior court found that Ross should not be spared the dea th penalty, and he was subsequently scheduled to be executed.
6. Kagan on Alice James: Kagan, 2000, pp. 96–97.
6. Alice James preferred instead to welcome death: Strouse, 1980.
7. Clones are typically easy to distinguish from the animals that donated their chromosomes: D. Baron's February 11, 2000 report, on National Public Radio (quoted at length in Chapter 19).
10. *At that point (1990), Lickliter had just published "The Phylogeny Fallacy: Developmental Psychology's Misapplication of Evolutionary Theory." This startlingly clear-sighted article contributed substantially to the effort to integrate developmental systems thinking with current theories in evolutionary biology.
13. Nematodes: Schaffner, 1998.
13. Fruit fly bodies are clearly influenced by nongenetic factors: Lewontin, 1983.
14. James on scientific theories: James, 1907, p. 46 (emphasis in original).

Chapter One

16. Words on the wall of The Pickerel: Izaak Walton, 1653.
17. Gottlieb on Caspar Friedrich Wolff: Gottlieb, 1992, p. 5.
17. *For ease of exposition, I am describing how this argument would have been leveled against an ovist, but the same argument could be used against a spermist, of course.
19. Gottlieb on Weismann's germ plasm theory: Gottlieb, 1992, pp. 60, 201.
19. "different . . . determinants entered different cells": Gilbert, 1994, p. 576.
19. Weismann acknowledges a certain "neo-preformationism": Weismann, 1894, p. 20.
20. The hypothetical "sequence on increasing perfection . . . was entirely static: Gould, 1977, p. 23.
20. Ontogeny recapitulates phylogeny: Haeckel, 1866, vol. 2, p. 300, cited in Gould, 1977, pp. 76–77.
20. "Phylogenesis is the mechanical cause of ontogenesis": Haeckel, 1874, p. 5.
20. *It is worth taking a moment to reflect on the unusual nature of this idea. Development occurs in the present (or, as we say in 2001, in "real time"). The evolutionary processes that produced our species, though, operated on our ancestors in the past. Haeckel's idea, then, effectively requires the past to somehow reach through time into the present, where it can effectively "direct" events in the here and now.
22. "To think that heredity will build organic beings without mechanical means is a piece of unscientific mysticism": His, 1888, p. 295.

247

22. "Every stage of development must be looked at as the physiological consequence of some preceding stage": His, 1888, p. 295.

22. The approach of the experimental embryologists: Gould, 1977, p. 194.

22. *It should be noted that Roux was one of Haeckel's students, making Roux's insistence that science move beyond Haeckel's ideas a remarkable instance of oedipal rebellion.

23. *Driesch, too, was one of Haeckel's students.

23. "Weismann's idea was only a little less crude than . . . preformation": Gottlieb, 1992, p. 60.

24. The fusion of several mouse embryos: Wolpert, 1991, p. 36.

27. "self-contained packets of inheritance": Gottlieb, 1992, p. 80.

27. Preformation being remembered more favorably: Cole and Cole, 1993, p. 82.

28. *Lewontin, Rose, and Kamin (1984) were making the same point when they wrote: "The floor area of a person's house and the average age to which he or she lives are positively correlated not because living in a big house is conducive to health but because both characteristics are a consequence of the same cause—high income." But they go even further in illustrating just how misleading correlations can be, by noting that "the distance of the Earth from Halley's comet and the price of fuel are negatively correlated in recent years because one has been decreasing while the other increased, but for totally independent reasons" (p. 99).

29. Morgan "set genetics on a course that diverged from embryology": Gilbert, 1992b, p. 211.

29. "Genetics and development went their separate ways": Gilbert, 1992b, p. 211.

29. Gottlieb on the genetics-embryology cataclysm: Gottlieb, 1992, p. 72.

30. The hostility between geneticists and embryologists: Gilbert, 1994, p. 37.

30. Genetics and embryology practiced on independent sciences: Gottlieb, 1992, p. 138.

31. "on the verge of a Renaissance of mammalian developmental genetics": Gilbert, 1992b, pp. 212–214.

31. Consequences of unification of the biological and behavioral sciences: Michel and Moore, 1995, p. 440.

Chapter Two

34. *Erasmus Darwin—Violetta's father and Charles and Francis's grandfather—was a noted British physiologist, naturalist, and poet who actually anticipated Charles's evolutionary ideas by some 60 years, albeit in a decidedly unfinished form.

34. Darwin's On the Origin of Species stimulated Galton's interest in inheritance: Kevles, 1995.

35. Galton concluded that "heredity governed . . . talent and character": Kevles, 1995, p. 4.

35. *A teacher named Richard Mulcaster used the words "nature" and "nurture" in 1582 to describe factors operating on the developing child (West and King, 1987). Plomin's (1994) identification of Galton as the first scientist to

use these words in this capacity reflects his estimation that Galton, unlike Mulcaster, was a scientist. For his part, Galton defined nurture in a stunningly vague way, as "food, clothing, education, or tradition . . . all these and similar influences whether known or unknown" (Galton, 1875, p. 9, cited in Gottlieb, 1992, p. 51).

35. Galton felt "justified in attempting to appraise [the] relative importance" of nature and nurture: Galton, 1883, p. 131, cited in Gottlieb, 1992, p. 50.

35. Galton's life's work involved devising methods to appraise the impact of nature and nurture: Gottlieb, 1992.

35. Galton on the value of twin studies: Galton, 1883, p. 155, cited in Gottlieb, 1992, p. 55.

35. "Nature is far stronger than Nurture": Galton, 1883, p. 168, cited in Gottlieb, 1992, p. 57.

35. Galton's earlier speculations: Galton, 1875, pp. 9–10, cited in Gottlieb, 1992, p. 51.

36. Galton's radical view about "a highly gifted race of men": cited in Kevles, 1995, p. 4.

36. Galton's eugenics: Kevles, 1995, p. 4.

37. "what Nature does blindly, slowly, and ruthlessly, man may do providently, quickly, and kindly": Kevles, 1995, p. 12.

37. International Health Exhibition in England: Kevles, 1995.

39. *I sometimes use scare quotes around the words "accounted for" to reflect my belief that while these are words used by behavior geneticists, they are nonetheless misleading. My reasons for believing this will become clear shortly.

39. The heritability of IQ is .70: Bouchard et al., 1990.

39. *Other types of studies that have been used to explore heredity include family studies and adoption studies; Billings, Beckwith, and Alper (1992) note that "other types of familial studies are subject to problems similar to those occurring in those using identical twins" (p. 229). Given that nontwin studies are subject to the sorts of problems that characterize twin studies *and* to other problems that do not characterize twin studies, I will focus only on heritability estimates generated using twin study data. These data are as good as it gets.

40. *Actually, this is true only if the environments in which identical twins are reared are no more similar than the environments in which fraternal twins are reared. This is an important issue; if individual fraternal twins are treated more differently than individual identical twins, differences between them could be accounted for by nongenetic differences just as easily as they could be accounted for by the fraternal twins' genetic differences. Behavior geneticists ordinarily deal with this problem by studying twins reared in adopted homes, because they believe that in this case, the environments of fraternal and identical twins can be assumed to be equally different. Whether or not this belief is valid remains a matter of debate, and I will consider the issue in more detail in the next chapter.

40. Plomin on the heights of fraternal and identical twins: Plomin, 1994, p. 44.

41. *This analogy has been on my mind intermittently since I first read about emergent phenomena in *Infancy*, the remarkable work of my exceptional mentors Jerome Kagan and Philip R. Zelazo (with Richard B. Kearsley, 1980, pp. 45–46).

42. Lewontin illustrates the heritability-causation difference using ordinary seeds: Lewontin, 1970.

44. *Behavior geneticists acknowledge that their studies do not explore the *causes* of traits in individuals and that, therefore, heritability estimates tell us nothing about the ability of environmental manipulations to influence trait development. But such acknowledgments still leave us with a major problem: statements like Plomin's about height—"most of the differences are due to genetic rather than environmental differences among individuals"—misleads many, if not most, readers. Those unfamiliar with the background knowledge that you now possess are likely to *interpret* such a statement to mean that a person's height is more influenced by her genes than by her environment, and such an interpretation is most definitely *not* accurate. Heritability estimates are about differences among those in a group and *not* about what causes an individual's traits.

44. Heritability reflects "*a particular population at a particular time*": Plomin, 1994, p. 43 (emphasis added).

44. *As Block (1995, p. 108) notes, "heritability is a population statistic just like birth rate or number of TVs and can be expected to change with changing circumstances. There is no reason to expect the heritability of IQ in India to be close to the heritability of IQ in Korea."

45. *How, then, was Galton able to use his statistics to estimate the likelihood that a parental trait would be "passed down" to offspring? Galton was successful because he made sure that the offspring in his studies were reared in the same environmental circumstances as were their parents. Fortunately, in a modern world in which we can manipulate environments so as to encourage positive developmental outcomes, there is little to be gained from understanding how "true" a particular human trait would breed in very narrowly specified environmental circumstances.

45. The *heritability* of number of fingers and toes in humans is almost certainly very *low*: Block, 1995, p. 103.

45. Opposable thumbs are not particularly heritable: Ariew, 1996.

45. Earring-wearing is highly heritable: Block, 1995, p. 104.

46. Behavior geneticists can simply ignore the effects of nongenetic factors: Block, 1995.

47. *While Galton's intellectual heirs in behavior genetics continue to actively pursue his vision, these days they use more sophisticated statistical techniques in their efforts. For example, if a particular environmental factor—say, exposure to racist treatment—is suspected of playing a role in the appearance of a trait—for example, low IQ—contemporary behavior geneticists might specifically measure the occurrence of the environmental

factor. Then, any correlation detected between the environmental factor and the trait could be subtracted from the heritability estimate; this makes sense in the present example, since such a correlation would mean that some of the measured variation in IQ is actually best accounted for by variation in exposure to racism. In this way, contemporary behavior geneticists might avoid generating obviously misleading conclusions about the influence of genetic factors on IQ.

Even so, the approach of today's behavior geneticists only helps when a particular environmental factor is *suspected* of playing a role in the appearance of a trait. If important environmental factors are not measured and evaluated with respect to their ability to account for variation in a trait, we are back where we started. In behavior genetics analyses, when genetic factors and environmental factors can *each* account for variation in a trait, if the environmental factors are not recognized, measured, and evaluated, the "credit" for the variation in the trait goes entirely to the genetic factors. That is, any variation in a trait not *specifically* accounted for by environmental factors remains available to be accounted for by genetic factors, even if unrecognized environmental factors contribute importantly to variation in the trait. The statistical methods used by behavior geneticists, then, are liable to produce inflated heritability estimates whenever they fail to recognize, measure, and evaluate important environmental factors. Any behavior geneticist unaware of how exposure to racism affects IQ would be at risk of drawing the misleading conclusion that in a racist society, variation in IQ is best accounted for by genetic factors.

Now, if we were very good at guessing in advance which environmental factors are important in influencing the appearance of which traits, this characteristic of behavior genetics analyses would not be a big problem. Unfortunately, as we will see, trait development is often impacted by subtle environmental factors in ways that are not at all obvious beforehand. And lamentably, we cannot take a wide-net approach either—measuring and evaluating the importance of *all* environmental factors—as there are just too many of them. But with no clear idea about which environmental factors must be measured to avoid inflated heritability estimates, such inflated estimates are unavoidable. Thus, aside from the fact that behavior genetics methods cannot tell us what *causes* traits in individuals, they also often lead us to think that direct genetic contributions to *differences* among individuals are larger than they in fact are. As a result, we can never be sure that some of the variation "accounted for" by genetic variation is not also accounted for by variation in environmental factors that were not studied.

48. *Quite recent advances in biochemistry and molecular genetics have raised the possibility of searching for the genetic origins of behaviors directly in DNA, but "the difficulties in demonstrating the genetic basis of these [psychological or behavioral] conditions using molecular methods are closely related to those affecting nonmolecular studies" (Billings, Beckwith, and Alper, 1992, p. 227). These researchers believe that DNA linkage methods

that try to identify specific genes that are directly influential in producing specific human behaviors are inevitably plagued by the same sorts of problems that plague traditional behavior genetics studies.

48. "identical twin studies offer no convincing evidence of the genetic basis of human behavior": Billings, Beckwith, and Alper, 1992, p. 228.

48. *For those to whom such things matter, I note that Billings and his colleagues generated this appraisal while working at an institution no less prestigious than the Department of Medicine and the Department of Microbiology and Molecular Genetics at Harvard Medical School.

Chapter Three

49. On Daphne Goodship and Barbara Herbert: Neimark, 1997.

49. On "squidging": Wright, 1995.

50. Half of all children in North America in 1900 had one of the ten most popular names for their sex: Schwegel, 1997.

50. *Keep in mind that even if we conclude that genetic similarity *is* the source of a particular trait's similarity in identical twins, that would not mean that the trait was caused by genes (or even that it was caused more by genes than by the environment). Since we are still discussing accounts of similarities and differences, the arguments presented in Chapter 2 remain valid: studying twins can *only* tell us about what causes *differences* (or similarities) in different people's traits, *not what causes the traits to appear as they do in the first place.*

51. Factors considered by adoption agencies in selective placement: Billings, Beckwith, and Alper, 1992, p. 229.

51. Twins "reared apart" are often raised by close relatives in similar environments: Michel and Moore, 1995, p. 207.

51. In one early study of twins reared apart: Shields, 1962.

51. "the most common pattern was for the biological mother to rear one of the twins": Lewontin, Rose, and Kamin, 1984, p. 107.

51. On twins Kenneth and Jerry: Lewontin, Rose, and Kamin, 1984, pp. 108–109.

52. *The fact that there are features that characterize *all* human developmental environments should not be misunderstood to mean that there are no important differences between cultures. The works of Rogoff and Morelli (1994) and Cole and Cole (1993) have made it clear that cultures do vary in ways that have important effects on development. But the existence of such variations in no way undermines the importance of the many similarities that characterize human environments in general.

52. *Some people might protest that environmental features required for the sustenance of life (for example, oxygen) should be ignored because they are not factors that actually *direct* development in any sort of specific way, but rather merely represent "background conditions" that support development. In later chapters, I will make the point that there is no good way to distinguish between factors that *shape* development and those that merely support it.

52. *Unlike individual laypersons trying to make sense of anecdotes about twins reared apart, professional twin researchers effectively sidestep problems of interpretation that arise from the existence of coincidences or from the fact that some features of human developmental environments are ubiquitous. These researchers are spared these concerns because their experimental design—which involves *comparing* fraternal and identical twins—takes care of these problems for them. Specifically, since coincidences and similarities born of ubiquity are just as likely among fraternal twins as among identical twins, any similarities seen among identical twins that are *greater than* those seen among fraternal twins must be due to the identical twins' identical genes. This approach, then, keeps researchers from erroneously attributing to genetic factors similarities that are really only coincidental or that in fact reflect similarities in twins' developmental environments. Thus, it is only when we *compare* identical twins reared apart to fraternal twins reared apart that we can have any hope of correctly attributing identical twins' similarities to their genetic similarities. Anecdotal evidence is of no value whatsoever in this endeavor.

55. *Considerable evidence now exists that a person's physical attractiveness influences in many ways the environment she or he encounters, from infancy through adulthood (Kleck, Richardson, and Ronald, 1974; Langlois, Ritter, Casey, and Sawin, 1995).

55. "Because identical twins are physically more similar than nonidentical twins, they will be treated more similarly even when reared apart": Michel and Moore, 1995, p. 207.

56. *Unfazed, behavior geneticists consider questions about the similarity of identical and fraternal twins' environments to be empirically researchable. As such, some studies have undertaken to scientifically *assess* the extent to which identical and fraternal twins reared apart experience different environments. Unfortunately, the results of these studies cannot adequately address the validity of the equal environments assumption, because it is impossible to guess in advance which environmental factors are important in influencing the appearance of particular traits. The problem with the empirical approach is that for such an approach to be successful, we would need to measure *all* environmental factors that could affect trait development (which, as far as we know, might include all environmental factors, period). The reason for this requirement is that any environmental factor *not* measured just *might* differ more for fraternal than for identical twins, and if such a factor impacts trait development, it could be counted on to produce greater differences among fraternal than among identical twins. These differences, then, would—in the presence of an unwarranted assumption—be erroneously attributed to genetic factors. At the moment, our understanding of which nongenetic factors contribute importantly to the appearance of traits remains quite limited. And in the absence of additional information, the only way to know for sure that twin studies meet the equal environments assumption would be to measure *all* environmental factors, a task that is not

practically possible. The equal environments assumption requires that *every environmental factor that could possibly influence a trait* be as similar for fraternal twins as it is for identical twins. But without information about every environmental factor that could possibly influence a trait, we cannot know if the equal environments assumption is warranted, leaving us at risk for violating it unwittingly.

57. *The placenta is a remarkable organ. In humans, it is co-constructed by the developing embryo and its mother; it is made of both maternal and embryonic cells that are "so intimately integrated that the two tissues cannot be separated without damage to both the mother and the developing fetus" (Gilbert, 1994, p. 238). During gestation, the placenta allows the mother to provide the developing embryo/fetus with both nutrients and oxygen and allows the developing embryo/fetus to eliminate its wastes by passing them on to the mother (who then eliminates them herself through the usual channels). Obviously, this organ is essential to successful development. Nonetheless, once the baby is born, this temporary organ is no longer needed, and so is expelled from the mother's body.

57. Placenta sharing leads to more similar IQs: Melnick, Myrianthopoulos, and Christian, 1978.

57. Placenta sharing leads to more similar personalities: Sokol et al., 1995.

57. *A more recent study (Riese, 1999) reported that the temperaments of *newborns* who developed a single chorion were no more similar to one another than were the temperaments of newborns who developed different chorions. Riese offers several possible explanations for the disparity between her observations and Sokol's observations; these include explanations based on methodological differences between her study and Sokol's, and explanations based on differences in maturational rates among newborns and young children. Sokol's data have not been called into question.

57. "Biological differences are associated with type of placentation": Beekmans et al., 1993, p. 548.

59. Gould on Chang and Eng: Gould, 1985, p. 69.

59. There are 1,031 cells in each male nematode's body: Schaffner, 1998.

60. Genetically identical worms reared in identical environments still differ: Gilbert and Jorgensen, 1998, p. 263.

60. *"developmental noise": Waddington, 1957. The discovery that organisms with identical genes do not develop identically even when they are reared in identical environments is certainly important. Nonetheless, lest we complacently shrug in the face of unexplained phenomena and simply attribute them to "developmental noise" without really understanding what causes them, Gottlieb cautions that the concept of developmental noise is essentially "non-analytic . . . [and therefore] not a useful idea for developmental analysis" (G. Gottlieb, personal communication, December, 1999).

60. A given trait is not predetermined, "even when the genotype and the environment are completely specified": Lewontin, 1983, p. 278.

60. "the number of eye facets differ" in a fruit fly's right and left eyes: Lewontin, 1983, p. 278.

60. *David Bowie, the renowned musician and actor, has had two different colored eyes ever since a schoolyard fight left him with a shiner as a boy.

62. Watson on lack of evidence for trait inheritance: Watson, 1930, pp. 103–104.

62. *In *Ontogeny and Phylogeny* (1977), Gould argued that major scientific theories typically *don't* fail *because* of a growing body of evidence that is incompatible with the theory. In the case of radical environmentalism, however, the evidence would have been even less important than usual; environmentalism would never have been rejected in the face of evidence from behavior genetics studies, because the logic of such studies does not permit the conclusion that genetic factors have causal roles in trait development.

Chapter Four

67. *In fact, this story *was* told by one Don Harper Mills, past president of a real organization called the American Academy (not Association) of Forensic Sciences, at that organization's 1987 banquet. While the story has attained urban-myth status as a result of its exposure both on the Internet and in *Magnolia*, in fact, Mills made up the story specifically to entertain his fellow forensic scientists.

68. Anderson on Sydney Barringer: From *Magnolia: The Shooting Script.*® Screenplay and Introduction copyright © 2000 by Paul Thomas Anderson. All other text, design, and compilation copyright © 2000 by Newmarket Press. Reprinted by permission of Newmarket Press, 18 East 48th Street, New York, NY, 10017.

69. *Might it be useful to lump together microenvironmental factors and genetic factors into a broad class of *biological* factors, whose influence on our traits could then be contrasted with that of the macroenvironment? Alternatively, might it be useful to lump together micro- and macroenvironmental factors into a broad class of *nongenetic* factors, whose influence on our traits could then be contrasted with that of the genes? Both of these approaches require initially determining if boundaries exist between biological factors and macroenvironmental factors, on one hand, and between genetic and nongenetic factors on the other. If we find that there *are* such boundaries, only then would we need to ask if making such distinctions might be helpful.

Does a clear boundary exist between genetic factors and nongenetic factors? If the microenvironment of a gene contained only nongenetic factors, these factors would be clearly distinguishable from the genes. However, in many important cases, the most significant factors in a gene's immediate environment are not nongenetic—they are other genes. If these genes, acting as the "environment" of another gene, did something different than nongenetic microenvironmental factors like hormones do, one might still hope to distinguish between genetic and nongenetic contributions to traits; unfortunately, as we will see, these types of factors do exactly the same sorts of things. Since microenvironmental factors like hormones have the same sorts of effects on genes as genes have on other genes, the distinction between genetic and microenvironmental factors is extremely blurry. Thus, attempts to separate genes and their microenvironments into different classes of influence are likely to produce only confusion.

How about the possibility that there is a discrete boundary between biological factors and factors outside of a person's body? This approach leads to confusion nearly as quickly as the first approach did. Most of us think of the environment as being "that which is outside of a body," because there seems to be an inviolable boundary between our environments and ourselves, namely our skin. But, as usual, a closer look reveals a more complex situation. In fact, we are all remarkably open to our environments. Each of us is constantly absorbing elements of the environment (air, water, food, light, vibrations, chemicals floating through the air, viruses, psychoactive substances such as caffeine—the list goes on and on). Simultaneously, the environment is constantly reclaiming elements of us (for example, our skin cells are continuously sloughed off into the air, forming dust in our homes; fat that seems permanently attached to our abdomens is metabolized as we take a late afternoon walk, some of the by-products of this metabolism exiting our bodies in our sweat and exhalation; the red blood cells constructed in the marrow of a woman's bones are flushed out of her body at the end of a month in which she did not conceive an embryo—this list goes on and on, too). The exchange between our bodies and our environment is uninterrupted and extensive, making the distinction between them blurry as well.

Perhaps we should not give up so easily, though. After all, we continue to find a distinction between daytime and nighttime to be useful, even though there is no sharp boundary between darkness and light; maybe we shouldn't let a little blur interfere with what we intuitively *know* is a good distinction. The quibbles I've raised notwithstanding, surely there is a valid distinction between that which is me and that which is my environment, even if there is a bit of blur at the boundary—no? But would such a distinction help us puzzle out the causes of our traits anyway? Many developmentists now believe that such a distinction, while defensible, does not help us to grapple with questions about the source of traits. These scientists argue that development is driven not by distinctive factors operating independently of one another but rather by the *co-action* of components that constitute a *single* integrated *system*. I will consider this viewpoint in detail in Chapter 10.

70. Psychologists are beginning to think in terms of cascades, too: Smith, 1999, p. 140.

70. Some philosophers would agree that it is unreasonable to call event A *the* cause of event Z: Mackie, 1965.

70. *The domino analogy used here—like all analogies—ultimately breaks down when applied to biological systems. It is not appropriate to think of

genetic factors as being analogous to event A or to think of the final production of a trait as being analogous to event Z, because in real biological
systems, *non*genetic factors sometimes initiate cascades of events that lead
to trait production via their influence on *genetic* activity. In later chapters, I
will spell out how this works in detail.

70. The cause of a murder is the murderer's behavior: Block, 1995.

70. *I discussed in the introduction my belief that the goal of science should be
to learn how to effectively *intervene* in natural events. I adopted this belief
after reading William James's arguments in his 1907 book, *Pragmatism.*

73. *I consider the definition of the word "information" on page 139. While the
best definition of this word remains the subject of some debate, I have chosen to use the definition favored through most of the twentieth century by
specialists in communications science: information is that which produces
one of a variety of possible states in a "receiver" (in this case, an animal)
because of the influence of a "sender" (in this case, either genetic or nongenetic factors). For more on this definition, see Johnston, 1987.

74. *There are three types of RNA involved in protein synthesis, respectively
called messenger RNA, transfer RNA, and ribosomal RNA, but as I'll only
be describing the role of one type, namely messenger RNA, I'll just refer to
it from here on as RNA.

74. "In humans, as little as ten percent of the DNA is translated": Maynard
Smith, 1993, p. 8.

74. "genomic light-years of noncoding DNA in between" our cistrons: Collins,
1999, p. 28.

75. *Because there are 20 amino acids that can serve as building blocks for proteins, a hypothetical protein that is only three amino acids long could have
any one of 8,000 possible shapes ($20 \times 20 \times 20$)! And because proteins are
often several *hundred* amino acids long, what this means is that the diversity
of protein shapes that is available is just enormous (if not infinite).

75. Protein shapes depend "on aspects of the intracellular environment": Johnston, 1987, p.160.

75. *Lewontin (2000) makes this point in slightly more technical detail. Referring to the fact that short chains of amino acids—called *polypeptides*—
become proteins when they fold up upon themselves into their characteristic shapes, he writes:

> *Of course, not all the information about protein structure is stored in the
> DNA sequence, because the folding of polypeptides into proteins is not completely
> specified by their amino acid sequences. That fact is conveniently ignored, because
> under the physiological conditions of normal cells the folding is unique. When cells
> are abnormal, however . . . different outcomes of the folding process may occur. . . .
> [T]he "correct" final structure of a protein . . . will not occur if the external
> conditions are not appropriate. We do not, in fact, know what the rules of protein
> folding are, so no one has ever succeeded in writing a computer program that will
> take the sequence of amino acids . . . and predict the folding of the molecule. Even
> programs that attempt very crude characterizations of the folding of regions of*

proteins . . . are not more than about 75% accurate. The difficulty is that a protein is not a string of amino acids, although it may be built up from them. It is a unique molecule with unique . . . properties that change during the process of partial folding. . . . Molecular biologists do not usually call attention to this ignorance about the determination of protein structure but instead repeat the mantra that DNA makes proteins (pp. 73–74).

75. A long route "between amino acid sequences and behavior": Johnston, 1987, p. 160.

75. *While it remains true that *most* of the details of trait development are still poorly understood, some headway has been made since Johnston wrote his brilliant 1987 critique of dichotomous approaches to development. Much of Parts II and III of this book consist of a presentation of some of what is now known about specific processes that lie on the developmental pathway between genetic factors and the traits they help produce.

75. Hair color is determined by the presence of melanin: Clayman, 1989.

75. *If melanin *was* a protein, the developmental route between genetic factors and the production of hair color would be shorter, but it would never be nonexistent. This should be obvious, given Johnston's insights about the influence of environmental factors on the shapes of proteins.

76. *One might protest that the effects of *abnormal* dietary intake of copper are not relevant to the normal development of hair color. I will address the importance of understanding abnormal development in a later chapter.

76. Hair colors of Himalayan rabbits: Cole and Cole, 1993.

76. *Temperature obviously affects hair color in normal, unplucked Himalayan rabbits as well. These rabbits' extremities are normally the only parts of their bodies covered with black fur because an animal's extremities are always the coldest parts of its body.

78. Dawkins's definition of a gene: Dawkins, 1976, pp. 30–34.

78. *According to Dawkins's definition of a gene, long strands of DNA—such as entire chromosomes—cannot be single genes. This is because before our chromosomes are stored into our sperms or eggs, they rearrange themselves via a process called "crossing over"; this process ensures that children are extremely unlikely to have *any* single chromosome that is identical to one in either of their parents' bodies. And according to Dawkins's definition, if a portion of a chromosome (or an entire chromosome) doesn't last for several generations, it cannot be a gene.

78. Definition of "cistron": *Penguin Dictionary of Biology.*

78. *Many biologists believe that introns do not code for anything at all, but Keller (2000, p. 61) reports that "the very distinction between introns and exons seems not to be fixed. In some cases, the synthesis of . . . proteins has now been shown to derive from stretches of intronic DNA—that is, from regions that had originally been consigned to the pile of 'junk' DNA, but, as we now know, mistakenly so." Perhaps such observations will ultimately help explain Lewontin's data (personal communication, February 21, 2001)

that indicate that natural selection operates on introns at least as powerfully as it operates on exons. Findings like these clearly suggest that introns cannot be appropriately thought of as "junk" DNA.

78. As much as 90 percent of a given cistron can consist of introns: Neumann-Held, 1998.

79. RNA splicing controls or regulates what "genes" do: Smith, Patton, and Nadal-Ginard, 1989.

80. Amara and colleagues' example of alternative RNA splicing: Amara et al., 1982.

80. Which molecule is produced depends on the type of cell doing the producing: Crenshaw, Russo, Swanson, and Rosenfeld, 1987.

80. Outcomes of DNA decoding do not "inhere in the DNA code": Michel and Moore, 1995, p. 224.

80. Alternative RNA processing is quite common: Neumann-Held, 1998.

80. One third of our DNA undergoes alternative splicing: Keller, 2000, p. 60.

80. *Alternative RNA splicing is not the only way in which a variety of different protein products can be obtained from a single DNA sequence. In an unbelievable—but not rare—process called "RNA editing," specific, individual nucleotide bases in an RNA sequence can be altered prior to protein production in a way that depends on the state of the developing organism at the time the editing occurs. As Neumann-Held (1998) reports, "There are two different kinds of mRNA editing. In one kind, nucleotides are inserted into or removed from the mRNA. The second kind converts nucleotides . . . [for example, A into G and vice versa]" (p. 122). As a result, given DNA sequences do not always give rise to the same protein products.

80. One gene, but hundreds of products: Keller, 2000, p. 61.

80. Neumann-Held on what is a gene: Neumann-Held, 1998, p. 125.

81. The gene concept is on "the verge of collapse": Keller, 2000, pp. 9, 69.

81. Genes "can only be functionally defined in a specific developmental context": Gray, 1992, p. 176.

81. "even today the gene is still . . . a hypothetical construct": Michel and Moore, 1995, p. 188.

Chapter Five

82. *In discussing the various rates at which animals develop, Gavin de Beer (1958) pointed out that toads metamorphose in early summer, that common frogs metamorphose in midsummer, and that bullfrogs metamorphose in their third year of life. He then noted that Julian Huxley—the elder brother of Aldous Huxley, who I discuss in Chapter 14—had previously shown that "differences in development of [these] animals can be explained on the basis of the different rates at which the thyroid gland develops in each. . . . The thyroid gland develops fastest in [the toad], which metamorphoses while it is still of small size, and slowest in [the bullfrog], which reaches a much larger size before it metamorphoses" (p. 26).

83. On the biochemistry of fertilization in sea urchins: Gilbert, 1992a.

83. Sea urchin sperm and eggs use the same biochemical pathway: Gilbert, 1992a, p. 483.

83. "Fertilization is the archetypal interaction between two cells": Gilbert, 1992a, p. 482.

84. Many developmentalists believe that interactions characterize normal development *in general*: Gottlieb, 1991a; Gottlieb, Wahlsten, and Lickliter, 1998; Griffiths and Gray, 1994; Johnston, 1987; Lickliter and Berry, 1990; Michel and Moore, 1995; Oyama, 1985; Smith, 1999; Thelen and Smith, 1994; Turkewitz, 1993.

84. *Mature cells of various types have different shapes, they produce and contain different collections of proteins (which contribute to their different shapes), and they perform different functions (as a result of their different shapes and protein contents).

85. Definition of "fate": *Random House College Dictionary*, 1988.

85. Wolpert on fate maps: Wolpert, 1992, p. 41.

87. "walking with parts of our body which we could have used for thinking": Excerpt from Spemann's autobiography, translated by Hall, 1988, p. 174.

88. *While a cell's neighbors constitute the *environment* of that cell, most people would consider such microenvironmental factors to be "biological," where this word refers to both genetic and nongenetic factors *within* a body. Ultimately, the portrait I will be painting of development will render these distinctions relatively unimportant, since the boundaries between the types of factors that influence development are significantly fuzzier than most people think, and since maintaining these distinctions turns out to impede efforts to understand development. For the moment, the important point is that factors *other than genes* (call them what you will, "biological," "microenvironmental," or just plain "environmental") have profoundly important roles in cell differentiation.

89. Smith on lessons from embryology: Smith, 1999, pp. 134–135.

Chapter Six

91. Sylvère's responses: Penfield and Jasper, 1954.

91. Penfield was surprised by reactions to his electrodes: Penfield, 1975, p. 27.

91. Sylvère was discharged in good condition: Penfield and Jasper, 1954.

92. *Surprisingly, an *adult's* cerebral cortex has only 100 billion cells, or 16 percent fewer than are present in a newborn's cortex. I will have more to say about this oddity shortly.

93. Sperry on neuronal ID tags: Sperry, 1965, p. 170.

95. Cortically "rewired" ferrets: Sur, Pallas, and Roe, 1990.

95. "Rewired" ferrets correctly interpret visual information as visual: Sur, 1993.

96. Deaf people use auditory cortex to process visual information: Neville, Schmidt, and Kutas, 1983, p. 127.

96. Blind people use visual cortex to determine a sound's location: Kujala, Alho, Paavilainen, Summala, and Näätänen, 1992.

96. Blind people use visual cortex to process touch information while reading Braille: Cohen et al., 1997.

96. *In addition, it is now known that the portion of the brain that receives sensory information from the right index finger (the reading finger) of blind Braille readers is significantly larger than the portion of the brain that receives sensory information from either the left index finger of these same people or from the right *or* left fingers of sighted, non-Braille readers (Pascual-Leone and Torres, 1993).

96. Cohen's conclusions: Cohen et al., 1997, p. 182.

96. The earlier the musical training, the greater the cortical reorganization: Pantev et al., 1998, p. 813.

97. *In a similar study of violinists, cellists, and a guitarist, Elbert and colleagues (1995) found that the extent of the brain's reaction to light touches of the fingertips depends on both musical training and on which hand's fingers are touched. Specifically, touching either nonmusicians' fingers or the fingers of the *right* hands of right-handed musicians generated less brain activity than touching the fingers of musicians' *left* hands. Why would this be? It turns out that right-handed players of stringed instruments finger their fretboards with their left hands while either bowing or strumming with their right hands; in general, then, the left hand's job involves considerably more manual dexterity than does the right hand's job. Consequently, Elbert et al. concluded that the most plausible explanation for their findings "is that the cortical territory [devoted to sensing] the left-hand digits . . . expanded" as a result of the musicians' experiences (p. 306).

97. Musical training affects the brain's *control* of the hands much as it affects the brain's *perception*: Amunts et al., 1997.

97. Cortical reorganization reflects "the pattern of sensory input processed by the subject during development": Pantev et al., 1998, p. 811.

97. *The reason higher level processes are likely to be at least as influenced by experience as are lower level processes is that the development of lower level processes—which are typically more essential to survival—is likely to be more canalized than the development of higher level processes. For those unfamiliar with the concept of canalization, I discuss it in detail in Chapter 13.

97. More connections in the human brain than stars in the Milky Way: Kandel, Schwartz, and Jessel, 1995.

98. *According to Kandel, Schwartz, and Jessel (1995), when "axons from the same eye . . . fire synchronously and thereby *cooperate* to . . . excite a target cell . . . this cooperative action on common target cells strengthens the cooperating axons at the expense of competing axons. . . . [In contrast, axons from different eyes] *compete* with each other . . . with the result that strong connections from one eye tend to inhibit the growth of the axons from the opposite eye" (p. 476).

99. Studies of kittens deprived from birth of stimulation in just one eye: Weisel and Hubel, 1965.

99. *It has recently become apparent that nature has an additional mechanism at her disposal to help create the basic structure of the brain: spontaneous neural activity. It turns out that in nonhuman primates, ocular dominance columns, like many other characteristic cortical structures, begin to form in utero; thus, the development of these features in these animals must be independent of their experiences with the macroenvironment. Katz and Shatz (1996) reviewed several studies of the spontaneous neural activity that occurs in brains prior to birth and concluded that such activity plays an important role in giving brains their structure. They wrote:

> *Early in development, internally generated spontaneous activity sculpts circuits on the basis of the brain's 'best guess' at the initial configuration of connections necessary for function and survival. With maturation of the sense organs, the developing brain relies less on spontaneous activity and increasingly on sensory experience. The sequential combination of spontaneously generated and experience-dependent neural activity endows the brain with an ongoing ability to accommodate to dynamically changing inputs during development and throughout life. (p. 1133)*

99. Experience-expectant processes: Greenough et al., 1987, p. 540.
99. Effects of experience on the development of species-typical brain structures: Johnson, 1997, p. 49.
100. Edelman on brain development: Edelman, 1992, p. 23; Edelman, 1992, p. 64.
100. Grafting limb buds onto chick embryos leads to more neuron survival: Hollyday and Hamburger, 1976.
100. Target tissue amounts do not affect initial proliferation of neurons: Hamburger, 1934.
102. *In a later chapter, when I discuss the effects of hormones on the brain, I will say more about how it is that proteins can serve as transportable, "informational" signals that can affect the activity of DNA.
102. *To date, a number of neurotrophic factors besides NGF have been discovered, and each seems to be responsible for inhibiting the suicidal tendencies of a different type of neuron. Collectively, trophic factors now appear to play several significant roles in the developing nervous system. In addition to influencing the survival of entire neurons, trophic factors have been implicated in the experience-based pruning of synapses (as happens when ocular dominance columns develop) and in the guidance of axons to their target cells (some trophic factors seem to be attractive to the growth cones of developing axons).
102. Studies of newborn kittens in visually impoverished environments: Blakemore and Cooper, 1970; Hirsch and Spinelli, 1970.
102. "the visual cortex may adjust itself . . . to . . . its visual experience": Blakemore and Cooper, 1970, p. 478.
103. *The fact that cells (and, as we shall see, genes themselves) are oftentimes *directed* by nongenetic factors should not be taken to mean that genes are unimportant. Genes are always as important as the environment in trait development; every trait you have reflects the activity of your genes. After

all, trophic factors—to choose just one example—could not be constructed without the contributions of the genes. But we ought not let the nature-nurture pendulum swing past vertical; *all* of your traits reflect genetic *and* nongenetic factors, as both types of factors are constantly co-acting to build the body (including the brain).

Chapter Seven

108. Temperature "determines" the developing turtle's sex: Bull, 1980.

108. *For other species of turtles, the details vary: female snapping turtles, for example, develop in eggs incubated at extreme temperatures (less than 20°C *or* greater than 30°C), while male snapping turtles develop in eggs incubated at more moderate temperatures. Temperature-based sex determination has the beneficial effect of reducing inbreeding: it ensures that brothers and sisters will not mate with one another, since all of the offspring in a given brood will normally be of the same sex. On the downside, such a mechanism also provides an easy path to extinction. If phenomena like an ice age or global warming occur, generations of same-sex animals could find themselves unable to reproduce for lack of available mates. If dinosaurs' sex determination was temperature-dependent, these sorts of environmental calamities could explain their sudden extinction (Ferguson and Joanen, 1982).

108. Coral reef fish change from female to male: Warner, 1984.

108. Hazardous linguistic shorthand is evident in Wolpert, 1992, p. 141.

109. "The only cells that are affected by the sex chromosomes are the germ cells": Wolpert, 1992, p. 138.

110. Jost sleeping with his rabbit subjects: Levine and Suzuki, 1998.

110. Jost's studies of sex determination in fetal rabbits: Jost, 1953.

110. *It should be noted that while testosterone is a so-called masculinizing hormone, it is nonetheless present in females as well; testosterone performs a variety of important functions in both male and female mammals.

111. *The ability of some proteins to regulate genetic activity explains how nerve growth factor can "inform" a neuron that it has built a functional synapse with its target cell, and so should be allowed to live. You'll recall that when a new functional synapse is formed, NGF is released by the target (postsynaptic) cell and subsequently absorbed by the presynaptic neuron and transported back to its nucleus. It turns out that NGF is a protein that can control the activity of some genes. Thus, when NGF enters the presynaptic neuron's nucleus, it interacts directly with that neuron's DNA, either inducing or inhibiting gene activity essential for the cell's survival or death. Either way, the NGF effectively "informs" the presynaptic cell that a functional synapse has been formed with the target cell. (NGF is not absorbed by presynaptic cells unless such a synapse is formed.) This is how proteins are able to carry functional signals to DNA, thereby influencing whether or not cells survive.

111. "We still do not know what the testis- or ovary-determining genes are doing": Gilbert, 1994, pp. 764–765.

111. People *with* an SRY gene can nonetheless develop female traits; people lacking an SRY gene can nonetheless develop male traits: Gilbert, 1994.

111. Androgen insensitivity syndrome: Meyer, Migeon, and Migeon, 1975.

111. *It turns out that the development of abnormal sexual characteristics is not extremely rare. Approximately one in 2,000 babies worldwide is born with ambiguous genitalia.

112. Social factors contribute to our genders, too: Bem, 1994; Maccoby and Jacklin, 1974.

112. Steroid-receptor complexes turn genes "on" or "off" or regulate decoding rates: see Yamamoto, 1985.

113. Sex steroids affect onset and growth of axons and dendrites: Toran-Allerand, 1976.

113. *In adolescence, when their atrophied ovaries fail to produce the estrogen they normally would, Turner's girls do not develop secondary sex characteristics such as breasts. Fortunately, treatment with estrogen and other ovarian hormones around the normal time of puberty can produce the physical and behavioral changes characteristic of this developmental stage.

113. Fetal feminization in response to maternal- or placenta-derived estrogen: Langman and Wilson, 1982.

114. Effects on fetal rat littermates of sharing maternal blood flow: Meisel and Ward, 1981.

114. "Stud Males and Dud Males": Clark, Tucker, and Galef, 1992.

114. Testosterone injected into a pregnant monkey's bloodstream can pass through the placenta and masculinize an XX fetus: Phoenix, 1974.

114. Fetal testosterone can cross the placenta and enter the mother's bloodstream: Meulenberg and Hofman, 1990.

114. *Even if a barrier were discovered that prevented the direct transfer of testosterone between twins, it might still be possible for a male twin fetus's testosterone to enter his mother's bloodstream and subsequently affect his twin sister, much as has been shown to occur in rats.

115. Can a human twin's hormones affect the other twin?: Resnick, Gottesman, and McGue, 1993.

115. *Remember, if we agree that it can be good to intervene in natural events, then we ignore at our own peril knowledge that could lead to helpful developmental interventions. Failing to study the effects of *normal* hormone exposure just because such exposure might "universally" characterize human development could keep us from discovering developmental interventions with the potential to improve lives.

115. The experiences of pregnant rats can affect their hormonal states *and* some of the sexual traits of their offspring: Ward, 1972; Ward and Weisz, 1980.

115. "the processes involved in masculinization . . . appear to have been compromised in the male fetuses of stressed mothers": Ward, 1972, p. 328.

115. Stressing pregnant rats reduces testosterone concentrations in male fetuses: Ward and Weisz, 1980.

115. Prenatal stress and human homosexuality: Ellis, Ames, Peckham, and Burke, 1988, p. 155.

115. *The methodological obstacles faced by these researchers prevented them from being able to rule out several other possible explanations for their findings. For instance, it is possible that mothers who experience the world as stressful may raise their sons in a way that influences their sexual orientation (M. Banerjee, personal communication, September, 1999).

115. *Those who imagine—erroneously—that fetuses are perfectly protected from the environment believe that all differences between newborns reflect genetic differences; if all fetuses experienced identical prenatal environments, genetic factors would be the only other possible source of their trait differences.

Chapter Eight

118. Schizophrenic women born to malnourished mothers: Susser and Lin, 1992.

118. The Dutch Hungerwinter effect in rats: Jones and Friedman, 1982.

119. "male offspring of underfed mothers deposit excessive amounts of fat": Jones and Friedman, 1982, p. 1519.

119. An experimental test of the insulin hypothesis: Jones and Dayries, 1990.

120. Insulin's effects on neurons: Recio-Pinto and Ishii, 1988.

120. Fetal insulin levels and the development of brain areas that regulate eating: Jones and Dayries, 1990, p. 1109.

120. Malnutrition and subsequent refeeding leads to higher insulin levels: Jones and Olster, 2000.

120. *It is convenient to caricature this fetus-environment interaction as a case in which the environment is *detected* by a fetus that then—somehow—alters *itself* accordingly. But a portrayal more in line with developmental systems theory would depict the environment as *contributing* to development by altering the fetus's development directly.

121. Fetuses sense stimuli in uterine environments: Macfarlane, 1978; Pedersen and Blass, 1982; Teicher and Blass, 1976, 1977.

121. *For example, scientists now know that a fetal rat's experience with the chemical environment it encounters in utero directly impacts its postnatal preferences for certain smells. Normally, after a female rat gives birth, the smell of her nipples elicits suckling from her newborn pups; we know this because washing the mother's nipples clean eliminates suckling (Teicher and Blass, 1976, 1977). Since coating a mother's washed nipples with amniotic fluid—the fluid that surrounds developing fetuses—elicits renewed suckling from her pups, Pedersen and Blass (1982) hypothesized that newborn pups might suckle specifically in response to the chemical "smells" they encountered in utero. To test this hypothesis, these researchers injected a lemon scent into the amniotic fluid surrounding fetal rats due to be born two days later. After the pups were born (and before they first attempted to suckle their mother), the mothers' nipples were washed and then perfumed with the same lemon scent to which the pups

had been exposed in utero. The study revealed that pups who were not exposed to the lemon scent in utero would not nurse from lemon-scented nipples under any circumstances. In contrast, newborn pups previously exposed to this scent *would* suck on lemon-scented nipples. Most remarkably, pups exposed to the lemon scent in utero were no longer willing to suck on unwashed nipples presumed to smell of ordinary amniotic fluid (which normally always elicit sucking). Pedersen and Blass concluded, "in mammals . . . prenatal events can influence the postnatal expression [of evolutionarily adaptive and organized behaviors]" (p. 354).

121. Incubator-reared mallard ducklings prefer the maternal assembly call *immediately* after they hatch: Gottlieb, 1965.

121. Gottlieb's research: Gottlieb, 1981.

122. *Daniel S. Lehrman made this general point in 1953 in a powerful critique of Konrad Lorenz's theory of instinctive behavior. In this devastating examination of the Nobel Prize–winning ethologist's work, Lehrman effectively argued that so-called isolation experiments—in which an animal is raised in isolation from other members of its species—do not allow for the identification of "instinctive" behaviors, Lorenz's assertions notwithstanding. Specifically, Lehrman argued that an organism can *never* be completely deprived of an environment, so that, at best, an isolation experiment can be used to illuminate what factors do *not* contribute to development, not what factors do contribute to development. He wrote:

> The isolation experiment . . . provides at best a negative indication that certain specified environmental factors probably are not directly involved in the genesis of a particular behavior. However, the isolation experiment by its very nature does not give a positive indication that behavior is "innate" or indeed any information at all about . . . [the] development of the behavior. (p. 343)

Moreover, Lehrman realized, because isolation studies rely on a scientist's intuitive sense about what sorts of experiential contributions to development are the important ones, they only explore *obvious* experiential contributions to development, not the sorts of environmental contributions that—while not as *obviously* consequential—might, in fact, be just as (or even more) significant. Lehrman concluded his denunciation of Lorenz's "instinct" concept with the following warning:

> Any instinct theory which regards "instinct" as immanent, preformed, inherited, or based on specific neural structures is bound to divert the investigation of behavior development from fundamental analysis. . . . Any such theory of "instinct" inevitably tends to short-circuit the scientist's investigation of [development]. (p. 359)

122. Mallard embryos breathe and vocalize a few days before hatching: Gottlieb, 1968.

123. Newborns are affected by in utero auditory stimulation: DeCasper and Spence, 1986.

123. Newborns choose their mother's voice over that of another baby's mother: DeCasper and Fifer, 1980.

123. The sucking method of studying newborn perception: DeCasper and Fifer, 1980.

124. Newborns might learn to prefer their mother's voice very quickly: DeCasper and Fifer, 1980, p. 1176.

124. Prenatal auditory experience affects nonhuman infants' preferences: DeCasper and Fifer, 1980, p. 1176.

124. Recordings of sounds in the uteruses of pregnant women about to give birth: Querleu and Renard, 1981.

124. *Turkewitz (1993) writes:

> During the course of gestation, the intrauterine environment undergoes changes consequent upon changes in the structure of the uterus. Thus, during early stages of pregnancy the uterus is relatively flaccid and thick-walled (somewhat like an underinflated balloon). In this condition it acts as a damper on externally generated sounds. As pregnancy proceeds, the uterine walls stretch, becoming thinner and tauter (in the manner of a slowly inflating balloon). When more distended, the uterus, like the fully inflated balloon, would be an amplifier rather than an attenuator of externally generated sound.

124. A more direct test of the prenatal experience hypothesis: DeCasper and Spence, 1986, p. 134.

125. DeCasper and Spence conclude that "the target stories were . . . preferred, because the infants had heard them before birth": DeCasper and Spence, 1986, p. 143.

125. *As usual, caution is warranted in trying to *apply* DeCasper and Spence's results to everyday parenting practices. Although development often entails a cascade of events so that normal experiences like hearing your mother's voice in utero might be important first steps in the normal sequence of events that leads to other, more "permanent," characteristics, we currently have no evidence that prenatal experiences of this sort necessarily have long-term consequences. In addition, we do not yet know how prenatal manipulations—like playing certain kinds of music loudly during the final weeks of pregnancy—could impact development; such manipulations could be detrimental, beneficial, or possibly even insignificant to the subsequent development of a person.

126. *It must be noted that Sperry and Gazzaniga's research subjects were people with epilepsy who had undergone neurosurgery to relieve the symptoms of their disorder. While the brain's right and left hemispheres do seem to be specialized for processing certain kinds of information, normal people have neurons that connect the two hemispheres, allowing these brain halves to share information *very* rapidly and to function in a fully integrated manner. The popular notion that some people are "left-brained" while others are "right-brained" is not consonant with what is known about normal brain function.

126. Newborns turn their eyes to the right when they hear speech, but toward the left when they hear nonspeech sounds: Hammer, 1977, cited in Turkewitz, 1993.

126. Newborns have a "right-ear ... bias for processing speech": Turkewitz, 1993, pp. 131–132.
126. Present-at-birth capacities need not be genetically determined: Turkewitz, 1993, pp. 125–126.
127. How verbal stimuli could come to be processed by the left hemisphere: Turkewitz, 1993, pp. 133–134.
128. Typical fetal posture could affect lateralization by causing the two ears to hear different sounds: Turkewitz, 1993, p. 134.

Chapter Nine

129. "increased beard growth is related to the resumption of sexual activity": Anonymous, 1970, p. 869.
130. Preventing newborn chicks from seeing their toes hinders mealworm consumption: Wallman, 1979.
130. *Note that a chick's normal response to mealworms is not "learned" in any sort of ordinary way, since experience with mealworms themselves is not necessary for the development of the response. Even so, this does *not* mean that the response develops independently of experiences; obviously, without specific kinds of experience, no response develops.
130. "everything the animal sees might influence ... its perceptual development": Wallman, 1979, p. 391.
131. Monkeys can develop a fear of snakes based on exposure to insects: Masataka, 1993.
131. *Masataka considered the following reactions to his stimuli to have been fear reactions: shaking or holding onto the side or back of the cage, vocalizing, or suddenly retreating to the back of the cage.
131. Development of snake phobia might depend on "wider perceptual experience with living beings": Masataka, 1993, p. 746.
132. Licking affects the sexual behavior of rat pups once they mature: Moore, 1992.
132. Understimulated rat pups exhibit deficient male sexual behavior: Moore, Dou, and Juraska, 1992.
132. Developmental experiences can affect mature behaviors in nonobvious ways: Gottlieb, 1991a.
133. *One of the reasons early experiences can be so important to trait development has to do with the cascading nature of biological events; given this characteristic, occurrences early in development can affect later outcomes in ways that are surprisingly far-reaching and sometimes quite unpredictable. This sort of phenomenon actually characterizes most dynamic systems, even those that are not biological. For example, in the 1960s, Edward Lorenz discovered the remarkable dependence of dynamic systems' final states on their early states while he was studying the development of weather systems. Lorenz's phenomenon came to be called "the butterfly effect," because incredibly small events—say, the fluttering of a butterfly's wings in Singapore—can theoretically affect the occurrence of small weather events nearby, which can affect the occurrence of even larger

weather events further away, and so on, ultimately contributing to the production of violent thunderstorms over Nebraska. Similarly, seemingly trivial events early in the development of biological systems can have significant consequences on later developmental outcomes.

133. *There are many sources of the idea that our early experiences indelibly affect the development of our traits. One of these sources is the Nobel Prize–winning work of Konrad Lorenz. Among Lorenz's most well-known concepts is the idea of "imprinting," in which a duckling or chick, upon hatching, follows the first moving object it encounters (its parent, in natural circumstances). Lorenz applied the notion of a "critical period" here, arguing that imprinting could occur only during a very specific period in the baby bird's life. Furthermore, he asserted that once acquired, the preference to follow the specific object would be permanent. Although the idea that there is a "critical period" for imprinting continues to be taught in many college courses, Michel and Moore (1995) have pointed out that "modern research does not support Lorenz's claims about the learning, critical periods, or irreversibility during imprinting" (p. 39).

133. Late developmental experiences can appreciably affect our traits: Bruer, 1999.

134. Male ring dove mating behaviors depend on testosterone: Michel and Moore, 1995.

134. Specific auditory and visual stimuli influence doves' hormone levels: Michel and Moore, 1995.

134. A mate can affect a dove's hormones indirectly, via feedback from the dove's behaviors: Michel and Moore, 1995, p. 159.

134. *Beauchamp, Doty, Moulton, and Mugford (1976) report on a variety of mammals that respond to chemical signals arising in other animals of the same species. While these authors argue that the traditional criteria used to classify substances as pheromones give rise to conceptual problems that render the pheromone concept of questionable value, their paper does not undermine the idea that stimuli arising in the macroenvironment can significantly influence an animal's hormone levels.

134. Pheromone-induced synchronization of women's menstrual cycles: Stern and McClintock, 1998.

134. Evidence of human pheromones: Stern and McClintock, 1998, p. 177.

135. "humans have the potential to communicate pheromonally": Stern and McClintock, 1998, p. 178.

135. A "pheromonal mechanism for menstrual synchrony": Stern and McClintock, 1998, p. 177.

135. *A word is in order about the identity of this intrepid scientist. Ultimately, this modern-day cross between Robinson Crusoe and Dr. Jekyll chose to suppress his identity and publish his data on the relationship between sexual activity and beard growth under the name of Anonymous, "for reasons," according to the journal's editor, "which may be self-evident" (p. 870). Nonetheless, the data, which were published in Europe's most prestigious English-language journal of science, *Nature*, were subjected to ordinary peer review and vouched for by

the anonymous scientist's colleagues. Thus, while the data leave us with all of the problems of interpretation associated with nonexperimental case studies done by scientists studying themselves, they are nevertheless interesting.

135. Increase in beard growth on resumption of sexual activity: Anonymous, 1970, p. 869.

136. Changes in beard growth just from being with a particular woman: Anonymous, 1970, p. 869.

136. Stress hormones can turn genes "on" or "off" or regulate their decoding: see Yamamoto, 1985.

136. Psychological stress affects immune response to agents of infectious illness: Glaser et al., 1987.

136. Immediate-early genes: Rosen et al., 1992, p. 5437.

137. "IEGs, are . . . an important link in the cellular machinery": Michel and Moore, 1995, p. 229.

138. IEGs, and how exposure to light affects mammals' circadian rhythms: Rusak, Robertson, Wisden, and Hunt, 1990.

138. Cells in hamsters' brains "undergo alterations in gene expression in response to . . . illumination": Rusak, Robertson, Wisden, and Hunt, 1990, p. 1237.

138. *Studies by Rusak, Robertson, Wisden, and Hunt (1990), of hamsters, taken together with Rosen et al. (1992) studies of kittens' IEG responses to light exposure, suggest that normal visual experience might affect brain development the way it does because such experience activates genes that regulate processes that contribute to normal brain maturation.

138. IEGs activated in response to canary and zebra finch birdsong: Mello, Vicario, and Clayton, 1992.

138. Songs produced by a bird's own species elicited the greatest IEG response: Mello, Vicario, and Clayton, 1992, p. 6821.

138. There is "a role for genomic responses in neural processes linked to song pattern recognition": Mello, Vicario, and Clayton, 1992, p. 6818.

139. Psychological stress can turn certain genes off: Glaser et al., 1990.

139. Stress-associated decrements in the human immune response "may be observed at the level of gene expression": Glaser et al., 1990, p. 707.

139. *This is the definition of information favored by Timothy Johnston, who discussed it in detail in his brilliant 1987 *Developmental Review* article "The Persistence of Dichotomies in the Study of Behavioral Development." Interested readers should consult this article for more information.

140. *Levels of the RNA studied by Mello et al. have been shown to increase during differentiation of neurons, suggesting that the cells producing this RNA were about to undergo a structural change.

141. Some deaf or blind people can recruit the auditory or visual cortex to process visual or auditory information, respectively: Cohen et al., 1997; Kujala et al., 1992; Neville et al., 1983.

141. Flexible responsiveness characterizes recovery after brain damage in adulthood: Kaas, 1991.

141. *I have decided to discuss the studies detailed here, because I believe that their implications are important enough to warrant presentation. I must be candid, though, and declare outright that I have serious concerns about the ethics of amputating healthy digits from any primate. Regardless of the benefits of this research to humanity, I remain saddened for the primates subjected to these experiments, and I look forward to the day when better technologies eliminate pressures to perform such destructive experimental manipulations.

142. *Note that these alterations also produce changes in the *sensitivity* of the remaining digits, as each remaining digit now has more brain cells devoted to processing the information it detects.

142. Adult monkeys' brains reorganize themselves after digit amputation: Merzenich et al., 1984.

142. Human amputees, too, experience functional reorganization of brain maps: Merzenich, 1998, p. 1062.

142. "basic features of . . . cortical maps . . . are *dynamically* maintained": Merzenich et al., 1984, p. 592 (italics added).

142. "the capacity to reorganize . . . may be a general feature of brain tissue": Kaas, 1991, p. 138.

142. "adult plasticity is a feature of all mammalian brains": Kaas, 1991, p. 161.

142. Kaas on how plasticity is possible in adulthood: Kaas, 1991, p. 162.

142. "differences in sexual behavior [can] cause . . . differences in brain structure": Breedlove, 1997, p. 801.

142. Experiential factors affect the number of *new* cells generated in *adult* rats' brains: Kempermann, Kuhn, and Gage, 1997.

143. Letting rats run supports survival of newly created neurons: van Praag, Kempermann, and Gage, 1999.

143. "the number of adult-generated neurons doubles . . . in response to training on . . . [specific types of] learning tasks": Gould et al., 1999, p. 260.

143. Experience affects the "proliferation and survival of newly formed neurons": Greenough, Cohen, and Juraska, 1999, p. 203.

143. New neurons can be formed in certain brain areas, even in human adults: Eriksson et al., 1998.

Chapter Ten

145. *The gene that codes for the protein that breaks down phenylalanine should not be thought of as "the gene for brown eyes," because plenty of people without PKU who *have* that gene nonetheless have blue eyes. Instead, this situation should drive home the central point of this book: even traits that we ordinarily think of as "genetic"—such as eye color—are in fact caused by interactions among a variety of genetic and nongenetic factors.

145. Textbooks say PKU is caused by a defective gene: see, for example, Cole and Cole, 1993.

145. *In the future, it might be possible to screen zygotes for the absence of the gene that codes for the protein that breaks down phenylalanine; in such a

scenario, treatment would involve altering the genome in only one cell. But for the moment, this is not yet an option.

146. PKU is the most common biochemical cause of mental retardation, and it afflicts one out of every 10,000 babies born each year: Diamond, Prevor, Callender, and Druin, 1997.

146. *Unfortunately, while children treated for PKU from birth typically score within the normal range on generalized tests of intelligence, they "often have significantly lower IQs than their siblings or other family members" (Diamond et al., 1997, p. 5). This suggests that the dietary treatment does not *completely* eliminate the effects of these children's genetic abnormalities. In addition, recent studies have "reported problems in attention control, concentration, problem solving, and 'executive functions'" among these children (Diamond et al., 1997, p. 5). Diamond and others have hypothesized that such findings reflect the fact that children unable to break down phenylalanine have reduced bloodstream levels of the *products* of normal phenylalanine breakdown; this occurs regardless of whether or not dietary intake of phenylalalnine is restricted. And since one of the products of phenylalanine breakdown is tyrosine—itself one of the "ingredients" required for the construction of the neurotransmitter dopamine—dietary interventions alone do not lead to the development of brains with perfectly normal chemistries. In addition, current tests of newborns' phenylalanine levels are not perfect, and so some at-risk newborns are missed; after a few months of consuming normal amounts of phenylalanine, most of these babies begin to develop the characteristic symptoms of PKU. While in most cases these symptoms cannot yet be reversed, recent research suggests that starting on a low phenylalanine diet can help even mentally retarded *adults* whose PKU was not detected in infancy. In particular, initiating dietary treatment of affected adults can lead to improvements in concentration, communication, and behavior. Given what we now know about the remarkable plasticity of the adult human brain, this finding is, perhaps, less than shocking.

148. Many developmental scientists hold that development is driven by the co-action of components of a single integrated *system*: Gottlieb, 1991a; Gottlieb, Wahlsten, and Lickliter, 1998; Griffiths and Gray, 1994; Johnston, 1987; Lickliter and Berry, 1990; Michel and Moore, 1995; Oyama, 1985; Smith, 1999; Thelen and Smith, 1994; Turkewitz, 1993.

149. Components of dynamic systems influence one another *bidirectionally*: Gottlieb, 1991a.

150. Studying walking in babies has shown the value of construing a person *and* environment as a *single* integrated dynamic system: Thelen and Ulrich, 1991.

150. Biological and environmental factors are necessary interacting *collaborators* in trait development; genes are just one developmental resource: Griffiths and Gray, 1994, pp. 277, 284.

151. Traits emerge from interactions between collaborating, integral components in a system: Gottlieb, Wahlsten, and Lickliter, 1998, p. 262.

151. No single component of the system contains *instructions* for building traits: Oyama, 1985.

151 *Unfortunately, the focus of this book precludes an adequately detailed explication of these ideas here, but interested readers can learn more by perusing books on the topic by Gleick (1987), Prigogine (1980), Stewart (1989), and Thelen and Smith (1994).

151. *The idea that natural systems can spontaneously organize themselves might seem somewhat radical, in that it appears to run counter to Sir Isaac Newton's second law of thermodynamics (the one that says that natural systems, *if isolated* and left to their own devices, get less and less organized with time). Fortunately, advances in the mathematics of chaos and in the study of certain types of nonlinear dynamic systems have shown that when systems are thermodynamically open—that is, able to "take in energy from their environment" (Thelen and Smith, 1998)—they are capable of temporarily becoming more organized over time. Again, a full explanation of this idea is, regrettably, beyond the scope of this book. Interested readers can refer for further information to Kelso (2000), Stewart (1989), and Thelen and Smith (1998).

152. Tornadoes have characteristic traits: Bluestein, 1999.

152. Thelen and Smith on thunderheads: Thelen and Smith, 1994, p. xix.

152. *An excellent description of the "development" of an ecosystem is contained in Chapter 2 of Lewontin (2000), *The Triple Helix: Gene, Organism, and Environment*.

152. "The brain is . . . a self-organizing system": Edelman, 1992, p. 25.

153. "the cause of development . . . is the relationship of the . . . components, not the components themselves": Gottlieb, 1991a, pp. 7–8.

153. "the question of [which component] is more important for . . . development is nonsensical because both are absolutely essential": Gottlieb, Wahlsten, and Lickliter, 1998, pp. 246–247.

154. The only way to "understand the origin of any [trait is] . . . to study its development in the individual": Gottlieb, Wahlsten, and Lickliter, 1998, p. 263.

154. Dormancy is also seen in animals: Clutter, 1978.

154. Lodgepole pinecones: Hackett, 1989, pp. 61, 73.

155. *Note, though, that a strict developmental systems theorist would not speak as if a pinecone somehow "detects" information. Instead, developmentalists would maintain that a fire sets in motion a causal chain of events that eventuates in the development of a pine tree.

156. Developmental systems theory being integrated into *all* of the life sciences: Gottlieb, Wahlsten, and Lickliter, 1998, p. 263.

156. A 1994 Nobel symposium organized around this theoretical framework: Magnusson, 2000.

Chapter Eleven

159. Long before Darwin, people considered how development is related to the "universal chain" of being: Gould, 1977.

160.   Development of the human heart: Gould, 1977.

160.   *In fact, it is now widely acknowledged that Haeckel misrepresented the extent to which some of his theoretical predictions had been supported by empirical observations.

160.   We imagine that natural selection delivers sets of genes to babies that determine the babies' traits: Lickliter and Berry, 1990, p. 357.

162.   *Wilson disease is a disorder of copper metabolism in which people with the gene "for" this disease cannot effectively eliminate from their bodies the small amounts of copper that we all ordinarily consume in our normal diets. As a result of this metabolic abnormality, excess copper accumulates in, and causes toxicity to, the liver and the brain, giving rise to liver disease and neurological or psychiatric disturbances such as depression, phobias, compulsive behaviors, aggression, and antisocial behavior. Despite the fact that Wilson disease is a so-called genetic disorder, it turns out that it is quite easy to treat; certain drugs (for example, penicillamine) remove copper from the tissues in which it accumulates, thereby allowing the excess copper to be eliminated from the body via urination. Of most importance, when people with the gene "for" Wilson disease are treated with these medications, they "can lead a perfectly normal life and . . . [be] indistinguishable from anyone else," despite the fact that each of their cells still contains the gene "for" Wilson disease (Lewontin, 1992, p. 30). Clearly, the gene "for" Wilson disease should not be seen as necessarily causing the symptoms of this disease.

162.   *Some might argue that bicycle pedals are "for" making the bicycle's wheels go around but that the ultimate effect of the wheels' revolution can vary in different contexts. Similarly, genes can be thought of as being "for" protein production, but the ultimate effects of their activity can vary in different contexts, too. I am not saying anything controversial here. As early as 1915, Sturtevant wrote, "Although there is little that we can say as to the nature of Mendelian genes, we do know that they are not 'determinants' in the Weismannian sense. . . . The difference between normal red eyes and colorless (white) ones in Drosophila is due to a difference in a single gene. Yet red is a very complex color, requiring the interaction of at least five (and probably of very many more) different genes for its production. . . . we can then, in no sense identify a given gene with the red color of the eye, even though there is a single gene differentiating it from the colorless eye . . . all that we mean when we speak of a gene for pink eyes is, a gene which differentiates a pink eyed fly from a normal one—not a gene which produces pink eyes per se, for the character pink eyes is dependent upon the action of many other genes" (Sturtevant, 1915, cited in Carlson, 1966, p. 69).

163.   *I used the words "effectively inherit" in this sentence because abilities can be "inherited" only in a very specific sense, and a very confusing one at that. Following Johannsen (1911), I argue below that we would be well served by thinking that traits per se—abilities included—cannot actually be *inherited* at all.

163. *"Phenotype" is defined in *The Penguin Dictionary of Biology* (1994) as the "total appearance of an organism, determined by interaction during development between its genetic constitution and the environment" (p. 476).

163. "natural selection favors . . . phenotypes, not genes": Mayr, 1963, p. 184.

163. *There has recently been a resurgence of interest in the inheritance of acquired characteristics among some biologists, particularly those interested in the activity of retrogenes in mammalian (and other vertebrate) immune systems. Those interested in these phenomena might wish to look into recent works by Barker (1993), Jablonka and Lamb (1995), and Steele, Lindley, and Blanden (1998). The latter authors write that in their book they use "current molecular knowledge of the immune system to show that there is scientific evidence consistent with . . . the inheritance of acquired characteristics" (p. 21).

163. *Darwin argued in the first edition of *On the Origin of Species* that the inheritance of acquired characteristics was compatible with natural selection but that it played an insignificant role in evolution relative to the role played by natural selection. Over the years, though, he changed his mind about the relative importance of such so-called Lamarckian mechanisms. In later editions of the *On the Origin of Species* and in other works produced later in his life, "Darwin gave an increasingly important role to Lamarckian mechanisms" (Barker, 1993).

164. Darwin's pangenesis hypothesis: Jablonka and Lamb, 1995, p. 7.

164. *Barker (1993) notes that reformulating Weismann's idea about the continuity of the germ plasm in more modern terms would involve asserting that the genes have causal influence over the appearance of traits but that the converse does not hold: the appearance of traits has no effect on the genes.

164. Black guinea pig pups can be born to a white guinea pig mother whose ovaries have been surgically replaced with those of a black female: Castle and Phillips, 1909, 1913.

165. *Note that the conceptual separation of development and evolution imposed by Weismann's doctrines is not the same as the disintegration of development and *genetics* that was discussed at the end of Chapter 1.

165. On the false idea that traits are determined *either* by events experienced during our lives *or* by factors that affected our ancestors before we were conceived: Lickliter and Berry, 1990.

165. Distinguishing *material* passed from parents to offspring and *traits determined* by that material: Michel and Moore, 1995.

166. "Heredity may then be defined as *the presence of identical genes in ancestors and descendants*": Johannsen, 1911, p. 159.

166. "as genetics increased in importance and influence, so did [Johannsen's] view of heredity": Jablonka and Lamb, 1995, p. 16.

167. The modern synthesis sees genes alone as "the material basis of evolutionary change": Jablonka and Lamb, 1995, pp. 16–17.

167. *You will recall that *all* of a trait's variation in a *population* can sometimes be "accounted for" by genetic factors even when that trait in an *individual* is

obviously caused by environmental factors as well; clearly, inferences about populations have no bearing on inferences about individuals.

168.   Traits can be reliably passed to offspring *only* if parents transmit both the genetic *and* environmental factors required to construct the traits: West and King, 1987.

168.   "an extended notion of inheritance is . . . a critical part of [developmental systems] theory": Griffiths and Gray, 1994, p. 283.

Chapter Twelve

169.   Studies of the effects on newborn mice of being "handled" by parents: Ressler, 1962, 1963, 1966.

169.   Amount of pup handling varies as a function of the *parents* doing the handling: Ressler, 1962.

170.   Black *or* white mice reared by white parents grew heavier and more exploratory: Ressler, 1963.

170.   "both strains . . . responded at a consistently higher rate if their *parents* had been raised by [white] . . . foster grandparents: Ressler, 1966, p. 266 (emphasis added).

170.   Environments provided by foster grandparents *for parents* affects the behavior of the "grandpups": Ressler, 1966, p. 267.

170.   Michel and Moore on Ressler's results: Michel and Moore, 1995, p. 51.

170.   "a nongenetic system of inheritance based upon the transmission of parental influences is potentially available to all mammals": Ressler, 1966, p. 267.

170.   *This was not an oddball, never-to-be-seen-again phenomenon. Denenberg and his colleagues (Denenberg and Whimbey, 1963; Denenberg and Rosenberg, 1967) reported that a similar effect can be pushed back a whole lifetime: by merely handling female rat pups for the first 20 days of their lives, Denenberg was able to affect the activity levels and weights of these rats' "grandpups" (this extends Ressler's findings another lifetime because Ressler's foster grandparents had never been subjected to any experimental manipulations themselves). Denenberg and Rosenberg (1967) concluded that "handling females in infancy can have an effect two generations further on" and that "the experiences which an animal has in early life will influence her unborn descendants two generations away by nongenetic mechanisms" (p. 550).

170.   Nongenetic inheritance underlying birds' species-typical songs, nesting sites, and nesting materials: Sterelny and Griffiths, 1999.

171.   Finches who sing their adoptive parents' song: Immelmann, 1969.

171.   "there can be a line of bullfinches that sing canary songs": Michel and Moore, 1995, p. 328.

171.   On the development of fire ant queens: Keller and Ross, 1993.

171.   Fire ant queens' development depends on pheromones produced in their colony of origin: Keller and Ross, 1993, pp. 121, 126.

172.   Developed trait differences "influence the future reproductive opportunities of [fire ant] queens": Keller and Ross, 1993, p. 127.

172. A fire ant colony's social organization "is perpetuated by virtue of the social environment in which new queens are reared": Keller and Ross, 1993, p. 121.

172. A transgenerational flow of "information" will occur "in any species in which learning . . . is important": Sterelny and Griffiths, 1999, p. 96.

172. *Any* time animals receive extensive parental care, they could acquire traits like those acquired by their parents: Michel and Moore, 1995.

173. Three types of "resources" organisms use as they develop: Griffiths and Gray, 1994.

173. *Note that Griffiths and Gray's threefold categorization scheme is not meant to be exhaustive.

173. *Since Griffiths and Gray consider genes to be a developmental resource that is fundamentally similar to other, *nongenetic* developmental resources, they consider genes per se to be parental resources.

173. *This notion was effectively captured by Gottlieb's concept of *the developmental manifold*, the emergence of which is described in a chapter in a forthcoming book by Lewkowicz and Lickliter.

174. Persistent environmental resources such as habitat can be acquired through evolution: Gottlieb, in press; West and King, 1987.

175. *Evelyn Fox Keller (2000, pp. 26–27) has commented on the inertness of genes, writing that, "left to its own devices, DNA cannot even copy itself: DNA replication will simply not proceed in the absence of the enzymes required to carry out the process. . . . Moreover, DNA is not intrinsically stable: its integrity is maintained by a panoply of proteins involved in forestalling or repairing copying mistakes, spontaneous breakage, and other kinds of damage incurred in the process of replication. Without this elaborate system of monitoring, proofreading, and repair, replication might proceed, but it would proceed sloppily, accumulating far too many errors to be consistent with the observed stability of hereditary phenomena."

175. Only in the *context* of "cellular machinery necessary for their functioning" will genes *do* anything at all: Oyama, 1992, p. 57.

175. Dennett refers to cellular machinery as a "reader" that interprets DNA: Dennett, 1995, pp. 113–114.

175. *Note that where Dennett uses the word "vicinity" in this quotation, he could just as correctly have used the word "environment."

175. The abnormal body plan of ether-exposed flies can be *inherited* through the *cytoplasm* of the egg: Ho, 1984.

176. Changes in the *nongenetic* materials in a zygote "can cause heritable variation": Sterelny and Griffiths, 1999.

176. "The material link between . . . development and evolution . . . realistically includes both cytoplasm and nuclear genes": Ho, 1984, p. 284.

177. Drawing distinctions between different types of developmental resources is unwarranted: Griffiths and Gray, 1994, p. 291.

177. Each type of developmental resource should be seen "as a *positive, informative,* and *constructive* force" in trait development: Lickliter and Berry, 1990, p. 358.

177. Behaviors can cause *environments* to be adapted to organisms: Odling-Smee, 1988, p. 77.

177. "The final step in the integration of developmental biology into evolution is to incorporate the organism as itself a *cause* of its own development": Lewontin, 1983, p. 279.

178. Adaptation can be caused by *both* natural selection operating on organisms and organisms operating on their environments: Odling-Smee, 1988, p. 77.

178. On the Galapagos woodpecker finch: Odling-Smee, 1988, pp. 76–77.

179. Children inherit both "a structured genome and . . . a host of other necessary influences and interactants as well": Lickliter and Berry, 1990, p. 355.

179. *What sorts of influences and interactants are passed—somehow or other—from generation to generation? Oyama's list of environmental factors that might perpetually be available during the development of successive generations includes "a maternal reproductive system, parental care or other interaction with [others of the same species], as well as relations with other aspects of the animate and inanimate world" (1992, p. 57). Michel and Moore's (1995) more detailed list includes cyclical phenomena such as seasons and day-night cycles, ubiquitous phenomena such as gravity, meteorological phenomena such as barometric pressure and ambient temperature, maternal, paternal, and sibling factors, and cultural factors such as the use of specific types of shelters and/or nests, foods, tools, and traditions (pp. 240–241).

179. "the full range of developmental resources represents a complex system that is replicated" during the development of successive generations: Griffiths and Gray, 1994, p. 277.

179. "an extended notion of inheritance": Griffiths and Gray, 1994, p. 283.

179. "elements of culture . . . are required for the replication of evolved psychological traits": Griffiths and Gray, 1994, p. 302.

Chapter Thirteen

181. Lamarck used the word "acquired" to mean "gain as a result of effort": Michel and Moore, 1995, p 177.

181. For Lamarck, a trait was "preserved by reproduction" when a descendant's trait resembled a trait found in its ancestor: Lamarck, 1809/1914, p. 113.

181. *Lamarck is well known—and uniformly ridiculed—by modern biologists for his idea that traits altered via one generation's efforts can be inherited by that generation's offspring. For example, Lamarck proposed that giraffes have long necks because each succeeding generation stretched its neck farther in order to reach leaves higher up in trees, and that the increased neck length acquired through the efforts of each generation was then "preserved by reproduction" in the next generation.

181. *This was not the first time the mutability of species had been asserted, to be sure—Charles Darwin's grandfather Erasmus, among others, had broached the possibility in earlier treatises—but Lamarck's version of the argument was so persuasive that nearly a century later Haeckel wrote:

> *Lamarck . . . is the real founder of this Theory of Descent . . . and it is a mistake to attribute its origin to [Charles] Darwin. Lamarck was the first to formulate the scientific theory of the natural origin of all organisms, including man, and at the same time to . . . [infer] from this theory . . . the descent of man from the mammal most clearly resembling man—the Ape. (1897, pp. 84–85, cited in Gottlieb, 1992, p. 17)*

Darwin himself, born in the year *Zoological Philosophy* was published, was a Lamarckist; he would no doubt have accepted Haeckel's contention that Lamarck deserved to be called the father of evolution. Darwin's contribution, of course, was *explaining* the mutability of species with his original concept of natural selection, no mean feat.

181. Weismann's narrow use of "inheritance of acquired characteristics": Jablonka and Lamb, 1995, p. 10.

182. *While there is only limited—and in some cases controversial—evidence that changes in somatic cells can lead to changes in germ cells, it has begun to look as if Weismann's barrier is *not* inviolable. Those interested in reading about this possibility can peruse Barker (1993), Jablonka and Lamb (1995), and Steele, Lindley, and Blanden (1998). It is possible, of course, that future research will confirm that some acquired characteristics can be inherited via penetration of Weismann's barrier; it seems unlikely, however, that such a mechanism will ever be considered to be of major evolutionary importance by mainstream biologists.

183. The traditional—and spurious—distinction between biological and "cultural" evolution "rests on a distinction between genetically 'inherited' and environmentally acquired traits": Griffiths and Gray, 1994, pp. 301, 302.

183. *Eunuchs are adult males who were castrated as boys. For example, in the Mughal Empire of fifteenth-, sixteenth-, and seventeenth-century India, young boys were sometimes stolen from their homes, castrated, and pressed into service as "hijra," whose "job" it was to look after the emperor's harem and to serve as his confidant. Even though these individuals were genetically normal—possessing both an X and a Y chromosome in every cell of their bodies—they nonetheless never developed normal male characteristics, due to the lack of testes-produced testosterone in their bodies. It is for this reason, of course, that powerful men trusted eunuchs with their harems.

183. If developmental processes show enough "stability over evolutionary time" that a trait develops in each new generation, the trait will be subjected to natural selection and be potentially "transmittable": Oyama, 1985, p. 102.

183. Evolutionary processes can lead to the reproduction of traits in descendants, regardless of the extent to which the traits depend on experiences for their development: Griffiths and Gray, 1994, p. 280.

184. All traits have the potential to be "transmitted" to offspring, just as all traits are acquired: Waddington, 1975, p. 45.

184. "There is no basis for the dichotomy of heritable-nonheritable because all traits arise as a consequence of . . . development": Gottlieb, 1992, p. 150.

184. *Note also that developmental systems theorists hold that some acquired traits can be reliably reproduced in successive generations *even if the DNA that contributes to the traits is never altered by any factor at all*. To a developmentalist, all that is required for such reproduction is stability across generations in the availability of the factors that cause the trait. Now, given the high-fidelity copying that characterizes DNA replication, such transgenerational stability is the rule in the case of genetic mutations; in contrast, such stability may or may not attend alterations to nongenetic developmental factors. Nonetheless, there will be times when such alterations *are* relatively stable, and in these cases, a novel trait acquired by a mother during her lifetime will be reproduced in—"passed down" to—her children. As long as traits can be adaptively altered during an individual's lifetime and then reproduced in successive generations, developmental systems theorists are satisfied that acquired traits can be "transmitted" to offspring; from their perspective, it doesn't really matter if experiences can alter DNA directly. Among the developmentalists who argue that such processes can underlie actual evolutionary changes are Baldwin (1896), Gottlieb (1992), and Waddington (1975).

184. *Many theorists would argue that these are all different forms of a single dichotomy. Oyama (1991) writes, "'nature vs. nurture' is less a dichotomy than a sprawling complex of multiply interconnected beliefs, metaphors, and associations. It is not easily disposed of, for it is deeply embedded in our thought, and it has as many conceptual relatives as it has guises. (Consider, for instance, the oppositions between essence and appearance, animals and humans, passion and reason.)" (p. 28).

185. Some behaviors resist experiential modification; other do not: Johnston, 1987, p. 178.

186. The word "innate" comprises a variety of subtly different notions: Oyama, 1991, p. 33.

186. *Similarly, it is simply not true that fetuses are unexposed to the environment (or, for that matter, that all fetuses develop in identical environments, owing to physiological controls over the characteristics of uteruses); recall Jones and Friedman's (1982) demonstration that—like human fetuses developing during the Dutch Hungerwinter—rat fetuses whose mothers are undernourished early in their pregnancies often develop into obese adults.

186. "Pointing out that a behavior is present at birth does not explain *how* it comes to be present at birth": Johnston, 1987, pp. 178–179.

187. Waddington defines "canalization": Waddington, 1975, p. 99.

187. Waddington portrays a trait's resistance to change as "a set of branching valleys": Waddington, 1975, pp. 258–259.

187. "Once a pathway is chosen it is entrenched or bound to produce a particular end state. It is this entrenchment that Waddington called canalization": Ariew, 1996.

188. *Many theorists who have used Waddington's idea in their attempts to explain behavioral or psychological development have written as if highly canalized traits are genetically determined and innate. Gottlieb (1991a) has

pointed out that canalization has appeared "in various developmental psychological models as a virtual synonym for what was previously called the *innate, native, or maturational* component in behavioral development" (p. 5).

188. Experiential canalization: Gottlieb, 1991b.

188. Gottlieb did not expect to be able "to demonstrate malleability in the presence of the species-specific maternal call": Gottlieb, 1997, p. 88.

189. Exposing ducklings to embryonic vocalizations buffers them "from becoming responsive to social signals from other species": Gottlieb, 1997, p. 88.

189. Canalization can arise from factors at *all* levels of developing systems: Gottlieb, 1991a.

189. Some theorists think we can tweak our understanding of "innateness" to capture the epigenetic qualities of development: Ariew, 1996.

189. Johannsen justifies his decision to coin the new word "gene": Johannsen, 1911, p.132.

190. *I touched upon this point earlier when introducing the notion of experience-expectant processes in connection with ocular dominance column development in a normal cat's visual cortex.

192. *Some dog breeds are known for their temperament, which in and of itself suggests that temperament development in these dogs is highly canalized. My conclusion that we cannot know how James's temperament would have developed if I had found him as a puppy reflects the fact that he was a mutt; information about what is typical for his "breed," then, is nonexistent.

192. "Heredity . . . becomes inseparable from development": Lickliter and Berry, 1990, p. 355.

Chapter Fourteen

193. *I am grateful that Stephen Jay Gould (1977) wrote about Huxley's book in *Ontogeny and Phylogeny*; I might not otherwise have read this fascinating and quirky novel.

194. The conclusion to *After Many a Summer Dies the Swan*: Huxley, 1939, pp. 353–355.

194. *In 1942, Julian Huxley published a book entitled *Evolution: The Modern Synthesis*, thereby giving a name to the new biological theory that he and a handful of other eminent biologists had collectively forged in the late 1930s. Even though he was directly involved with the origination of what is essentially a nondevelopmental theory, "Huxley was one of the few architects of the modern synthesis to even mention changes in ontogeny as contributory to the new outlook on the causes of evolution" (Gottlieb, 1992, pp. 125–126). Aldous and Julian's interest in biology and evolution was not surprising: their grandfather was the famous zoologist Thomas Huxley, the foremost supporter in nineteenth-century England of Charles Darwin's theory of evolution.

195. "evolutionary changes must appear in ontogeny": Gould, 1977, p. 214.

195. "[Evolution] has never been a direct succession of adult forms, but a succession of ontogenies": Garstang, 1922, p. 82.

197. Heterochrony: de Beer, 1958, p. 34.

197. Evolutionary modification involves "a change of timing for developmental stages already present in ancestors": Gould, 1977, p. 214.

197. The Irish elk: Gould, 1977.

198. Two ways traits of adult ancestors can come to characterize juvenile descendants: Gould, 1977.

198. Two ways traits of juvenile ancestors can come to characterize adult descendants: Gould, 1977.

198. Neotenous traits in domesticated dogs: Coren, 2000.

199. On domestication of the Russian silver fox: Trut, 1999.

199. "breeding for tameability . . . has resulted in [adult] dogs that are . . . more like wolf puppies": Coren, 2000, p. 213.

199. Neoteny in humans: Gottlieb, 1992, p. 100.

199. "human beings are 'essentially' neotenous": Gould, 1977, p. 365.

199. *It is here that we find an idea that so struck Aldous Huxley that he constructed an entire novel around it: If human beings are fetal apes, then, in the mind of a novelist, a concoction of fish guts that prolongs life (and development) could conceivably help a human who ingests it to ultimately develop full-blown mature apelike characteristics.

199. "Ontogeny does not recapitulate Phylogeny: it creates it": Garstang, 1922, p. 98.

200. "the first Bird was hatched from a Reptile's egg": Garstang, 1922, p. 99.

200. *If you don't yet see how Garstang came to this conclusion, think of it like this: All chickens must come from eggs, but not all eggs must come from chickens (since reptiles lay eggs, too). Therefore, the egg came first.

200. Both Garstang and de Beer thought evolution results only from changes in genes: Gottlieb, 1992, p. 174.

200. "there is so much untapped potential in the existing development system . . . that evolution can occur without changing the genetic constitution of a population": Gottlieb, 1992, p. 174.

200. *Gottlieb's (1992) book ends with speculation that anatomical changes could conceivably appear even as an effect of novel behaviors that are *not* produced in response to novel environments.

201. Novel developmental circumstances can even lead to anatomical changes: Gottlieb, 1992, pp. 176–177.

201. "a change in the developmental conditions activates heretofore quiescent genes, thus changing the usual developmental process and resulting in an altered . . . phenotype": Gottlieb, 1992, p. 193.

201. Evolutionary changes "need not involve loss of genetic information": Kollar and Fisher, 1980, p. 995.

202. When development occurs after extreme climate changes, radical anatomical novelties can emerge: Rutherford and Lindquist, 1998, p. 336.

202. Some of Rutherford and Lindquist's flies' deformities could be favored by natural selection: Cossins, 1998.

202. "a plausible mechanism for promoting evolutionary change in otherwise entrenched developmental processes": Rutherford and Lindquist, 1998, p. 336.

202. "populations contain . . . unexpressed genetic variation that is capable of affecting . . . traits. Sometimes very specific [environmental] conditions can uncover this . . . variation": Rutherford and Lindquist, 1998, p. 341.

203 Troops of Japanese monkeys who receive food offerings from tourists: Michel and Moore, 1995, p. 161.

203. New data appear to render Gottlieb's hypothesis plausible: Rutherford and Lindquist, 1998.

204. On the failure of the modern synthesis to deal adequately with development: Waddington, 1975, p. 280.

Chapter Fifteen

207. *In *Not in Our Genes: Biology, Ideology, and Human Nature* (1984), Lewontin, Rose, and Kamin took an unabashed stance on the political left when discussing these issues; another book by Lewontin, *Biology as Ideology: The Doctrine of DNA* (1991), takes this position as well. Similarly, the ideas advanced by Richard Dawkins in his wonderfully stimulating *The Selfish Gene* (1976) and by Herrnstein and Murray in *The Bell Curve* (1994) were self-consciously supportive of positions on the political right. In contrast, I have endeavored to write a book that is less political in its orientation, because I believe wholeheartedly that the public is better served by this sort of presentation of scientific facts. Now, I am aware that even "facts," such as they are, cannot help but reflect the social and political contexts in which they are first conceived. In addition, *any* fact is liable to have political implications, implications that, in many cases, will be apparent to the writer *before* the writer—who cannot be completely without political opinion—actually gets the "fact" on paper. As a result, the process of self-editing, be it conscious or unconscious, renders it unlikely that there can be any such thing as a politics-free presentation of politics-free facts. Nonetheless, my presentation of the facts of development as I (and many others) see them has not been *motivated* by any political agenda of which I am consciously aware.

208. The goal of discovering the sequence of the human genome by 2005: Collins, 1999.

209. Lewontin reconciles his skepticism of the human genome project with the fact that it has attracted lots of talent and cash: Lewontin, 1991, pp. 51–52.

210 The "range of reaction" idea: Platt and Sanislow, 1988.

210. "a genotype determines an indefinite but circumscribed assortment of phenotypes, each of which corresponds to one . . . possible environment": Gottesman, 1963, p. 254.

211. "we can never be sure that any of these traits have reached the maximal development possible with a given genotype": Dobzhansky, 1955, p. 77.

211. "new environments are constantly produced": Dobzhansky, 1955, p. 75.

212. "One genotype may grow better than a second at a low temperature, but more poorly at a high temperature": Lewontin, Rose, and Kamin, 1984, pp. 268–269.
213. Transplantation experiments with five-day-old chick embryos: Kollar and Fisher, 1980.
213. Chick genes can contribute to teeth with "perfectly formed crowns": Kollar and Fisher, 1980, p. 993.
213. A bird's normal toothlessness is "not a consequence of . . . genetic coding. Rather, an upset of a developmental sequence . . . [blocks] tooth development": Kollar and Fisher, 1980, p. 995.
213. On the possibility of cows with wings: Michel and Moore, 1995, p. 185.
214. *It is, of course, theoretically possible—even likely—that genes specify a range of developmental outcomes, but since the limits of that range can never be known, this possibility is of absolutely no practical consequence at all.
214. *While the ability to clone complex animals from adults in a laboratory is a recent technological advance, it has been possible for many years to study clones by examining the genetically identical offspring produced naturally by asexually reproducing plants and animals.
214. *I cannot *begin* to imagine such an environment, of course, and it is likely that a vanishingly small number of environments would produce this outcome, but these facts alone do not diminish the validity of this argument. If *any* developmental environment could produce this outcome, it cannot be argued that Mr. Jordan's genes are *inherently* structured to produce a superior basketball player.

Chapter Sixteen

217. Evidence contradicting the claim that black Americans score lower on IQ tests: Tate and Gibson, 1980.
217. "Blacks, on the average, suffer from less access to health care, obtain less prenatal care, and live in more impoverished and stressful residential areas": Fairchild, 1991, p. 104.
217. Black babies are almost three times more likely . . . to be born with low birth weights: Lieberman, 1995.
218. "There is no reason to expect the heritability of IQ in India to be close to the heritability of IQ in Korea": Block, 1995, p. 108.
218. *Moreover, any such changes will be *independent* of the IQ scores that would have been obtained if development were allowed to proceed in the present environment. That is, those individuals who develop inferior IQ scores in the present environment could develop inferior, superior, or commensurate scores in a different environment.
219. When poor children are adopted into wealthier families, their IQ scores are ultimately much higher than their biological parents' scores: Lewontin, 1991.
219. "IQ in many countries has gone up 15 points, about the same as the gap separating blacks and whites in the USA": Block, 1995, p. 106.

219. *This conclusion was reached in Chapter 2 via Lewontin's example. To recap, if genetically diverse seeds are allowed to develop in a uniform environment, *all* of the height variation seen among the mature plants will be "explainable" solely in terms of the genetic diversity originally present in the seeds. The heritability of height in this environment will be 100 percent since the variation in the plants' heights cannot reflect variations in environmental factors at all, as all environmental factors in this example are held constant. But given our knowledge that environmental factors (for example, light, water, and nutrients) affect the development of plants' heights in important ways, affecting this trait with an environmental manipulation would be patently simple, the enormous heritability of height notwithstanding: enriching the developmental environments of these seeds would result in taller plants.

220. *We must keep in mind, though, that high heritability in a trait does not mean that children will *inevitably* develop the trait if their parents did. If children develop in an environment that is different in important ways from the environment in which their parents developed, it is entirely possible that they will not develop the trait of concern, regardless of its heritability. It is also worth considering whether or not parents would really have any use for estimates of the heritability of traits like IQ anyway: Would anyone ever say "I love him, but I'm afraid of having children that are as stupid as he is"? It seems unlikely that heritability estimates for most psychological characteristics would be of value to individuals making decisions about procreation.

220. Labeling a trait "heritable" often stops investigation of the trait's *development:* Lehrman, 1953, p. 359.

220. *In addition to the obvious ethical issues raised by such manipulations, at the moment, we would have *no idea at all* about where to begin making such alterations in the genome.

221. *The reality of the resistance to these ideas is easily demonstrated. Remember, an astute observer could have seen the coming failure of the genes-cause-traits idea as early as 1891, when Driesch conclusively demonstrated the equipotentiality of all early embryonic cells. Despite this demonstration, the genes-cause-traits idea persisted through the next century, certainly reflecting resistance of some sort.

221. "There may be genuine scientific reasons for suspecting that the synthetic theory does not offer a complete explanation of evolution": Bowler, 1984, p. 342.

Chapter Seventeen

222. The synthetic theory is attractively simplistic: Odling-Smee, 1988, p. 75.

223. "biological explanations . . . were used to . . . justify the inferior social status of women and minority ethnic groups": Alper and Natowicz, 1992, p. 666.

223. Biological explanations often reflect "values associated with the more politically powerful groups in society": Michel and Moore, 1995, p. 44.

223. Less politically powerful groups now using biological explanations for their own traits: Maddox, 1991.

223. Explanations for why less powerful minorities are now using biological explanations for their own traits: Alper and Natowicz, 1992.

223. Biological explanations might help society see some behaviors as uncontrollable; some hope this "will have the positive social consequence of removing blame and discrimination": Michel and Moore, 1995, p. 44.

225. *I risk repetitiveness here, because the following point is so important. Even when differences among people's traits are *completely* accounted for by differences among their genes, this still does not mean that the traits themselves are unaffected by nongenetic factors during development. The presence of certain genes in a person's body can increase the likelihood that that person will develop particular behaviors; *nonetheless, these behaviors will actually emerge only in response to experience with particular developmental circumstances.*

225. Psychological interventions affect our brain chemistries just as medical interventions affect our psychological processes: see, for example, Baxter et al., 1992, and Schwartz et al., 1996.

226. The Federal Violence Initiative: Wright, 1995.

226. Goodwin says biological factors "inclining human beings toward violence . . . might be [used as] 'biological markers' of violent disposition": Wright, 1995, p. 69.

227. On the multiple causes underlying the emergence of violence: Guerra, 1994, p. 661.

227. *Those with a developmental systems perspective would maintain that genetic factors, too, must contribute to the appearance of violent behaviors, because genetic factors contribute to the development of all of our traits. Nonetheless, any attempts to reduce the violence that currently afflicts our communities will necessarily involve manipulations of environmental factors, for both technological and ethical reasons.

227. Violence and delinquency prevention programs that provide mothers with prenatal care and children with preschool enrichment experiences: Zigler, Taussig, and Black, 1992.

227. Curtailing aggression by teaching kids how to resolve conflicts nonviolently: Guerra, 1994.

227. *As Guerra (1994) notes, there are various macrosocial factors that, if they remain stable, might impede efforts to reduce violence in our society; these include "poverty, economic inequality, overcrowded housing, inadequate services, racism, discrimination, media portrayals of violence, and the availability of guns" (p. 663). Clearly, we have our work cut out for us.

Chapter Eighteen

228. *Note that studies of population genetics (like studies of behavior genetics, which use related methods) are the original source of many of the media claims that genes "for" specific traits—both normal and abnormal—have been discovered.

228. Crick on "bean-bag genetics": Dennett, 1995, pp. 101–102. It should be noted that Ernst Mayr coined the phrase "bean-bag genetics" much earlier, in 1959, to describe the methodological approach of population genetics.

229. *According to some estimates, as many as one out of every ten people of north European descent carry at least one copy of the mutated gene associated with hemochromatosis (Burke et al., 1998).

229. Consequences of hemochromatosis: Burke et al., 1998.

229. *The standard treatment for hemochromatosis is called "phlebotomy." This scary word is fancy doctor-speak for taking blood. While one might argue about whether this name is any less frightening than the name for the identical medieval treatment—"bleeding" a patient—in fact, the treatment is exactly the same as if a person were required to "give blood" to the Red Cross a few times every year. Periodically taking blood from patients with hemochromatosis effectively removes enough of the excess iron from their bodies that they are subsequently able to lead otherwise normal lives.

229. "genetic testing is not recommended at this time . . . for . . . hemochromatosis": Burke et al., 1998, p. 172.

229. "simple detection of the mutation does not predict the most likely clinical course" of hemochromatosis: Collins, 1999, p. 32.

230. Among those homozygotic for hemochromatosis, symptoms range from "none that are detectable" to "severe organ damage from iron overload": Collins, 1999, p. 32.

230. On "the need for caution in the use of genetic testing": Burke et al., 1998, p. 176.

230. Genetic mutations associated with PKU do not *invariably* produce PKU: Burke et al., 1998.

230. "there may always be an inherent uncertainty in a hemochromatosis diagnosis made on the basis of genotype": Burke et al., 1998, p. 176.

230. "variable [genetic] expressivity . . . is the rule rather than the exception": Burke et al., 1998, p. 176.

231. "a phenotypic screening test is always preferable to a [genetic] test": Burke et al., 1998, p. 176.

231. *This disease has previously been controlled only by forcing afflicted infants to live out the first year of their lives in sterile plastic bubbles where they are protected from infections that they cannot fight off with their compromised immune systems. In cases where suitable bone marrow donors have been found when the children are older, transplants have then been attempted.

231. On "bubble babies" being "cured" by gene therapy: Maugh, 2000.

232. Gene therapy "has proven to be a much more complex undertaking than was initially imagined": Maugh, 2000.

233. *It is *this* phenomenon, and not some characteristic of my genes per se, that leaves me rather confident that I will not grow wings in my lifetime. By the same token, I suspect that I will never be a world-class basketball player, not

because my genes necessarily constrain my competence—perhaps my genes, in combination with *some* developmental environment, could have produced an agile 7-foot, 2-inch jumper with a soft touch—but because the developmental pathway that I have been headed down for some time now appears to be taking me ever farther from that endpoint. Then again, this is exactly the sort of argument that would have been offered less than five years ago to support the claim that cloning an animal from a mature, differentiated cell was impossible. For many years now, textbooks of developmental biology have taught that "the progressive restriction of . . . potency during development appears to be the general rule" (Gilbert, 1994, p. 42). Despite this general rule, five years ago, Ian Wilmut was able for the first time ever to return the nucleus of a mature, differentiated cell—one that had already traveled a long way down a normal developmental pathway—to an immature, undifferentiated, and therefore potent state. He did this by altering the *environment* of the adult cell; the upshot of this novel manipulation was the development of Dolly the sheep, the first mammal cloned from an adult cell. For the moment, though, development of *whole organisms* seems to always proceed in one direction—a direction that entails the narrowing of potential outcomes—leaving me earthbound with neither wings nor a Jordan-powered jump.

233. Methods to assess the vision of preverbal infants: Fantz, Ordy, and Udelf, 1962; Dobson and Teller, 1978.

234. *Given Kollar and Fisher's (1980) demonstration that the undifferentiated cells of bird embryos can be induced to differentiate into mammalian teeth, it is apparent that, in some cases, undifferentiated cells even have the potential to become cell types normally found in other species!

234 The first scientists to successfully culture embryonic stem (ES) cells in a laboratory dish: Thomson et al., 1998.

234. "ES-derived cells have been successfully transplanted into . . . mice": Gearhart, 1998, p. 1061.

234. "using existing organs as scaffolding, replacing the original cells with those derived from ES cells": Solter and Gearhart, 1999, p. 1468.

234. Implanting fetal cells into the substantia nigra of a Parkinson's disease sufferer: Lindvall et al., 1990.

234. Implanted cells restored neurotransmitter synthesis in the patient's brain: Lindvall et al., 1990, p. 547.

234. Transplanting fetal cells is controversial for several reasons: Stein and Glasier, 1995.

235. *Note that the work of the Swedish-English-Swiss team reported here was completed several years before the Wisconsin team successfully cultured ES cells in vitro (Thomson et al., 1998). ES cells were not available to the Swedish-English-Swiss team when they first attempted their innovative treatment of Parkinson's disease in the late 1980s.

235. Transplanting fetal cells is ethically questionable: Crutcher, 1995, p. 54.

235. We can imagine a woman conceiving just to save her father from Parkinson's disease: Sanberg, 1990.

Chapter Nineteen

240. *Given the relatively small number of macroenvironmental factors involved in the development of coloration relative to the number of such factors involved in the development of the brain structures and functions responsible for personality, we should expect personality variations to be at least as common among clones as are variations in coloration.

Epilogue

242. Simone de Beauvoir had it right—"a human being is . . . the being whose essence is in not having an essence": Lewontin, 1991, p. 123.

242. Prominent geneticists still make developmentally nonsensical statements: Jasper Rine of the University of California, Berkeley, quoted in Mann, 1994, p. 1687.

243. On the notion that XYY men are more aggressive than normal (XY) men: Mann, 1994, p. 1687.

243. Journalists still report not-yet-proven associations between traits and genes: Mann, 1994, p. 1687.

243. Behavior geneticists and journalists need to be aware of the developmental systems perspective: Gilbert and Jorgensen, 1998, p. 265.

244. Genes are "a cultural icon, a symbol, almost a magical force . . . a secular equivalent of the Christian soul": Nelkin and Lindee, 1995, p. 2

244. Genetic essentialism "reduces the self to a molecular entity, equating human beings . . . with their genes": Nelkin and Lindee, 1995, p. 2.

REFERENCES

Alper, J. S., and Natowicz, M. R. 1992. The allure of genetic explanations. *British Medical Journal* 305:665–668.

Amara, S. G., Jonas, V., Rosenfeld, M. B., Ong, E. S., and Evans, R. M. 1982. Alternative RNA processing in calcitonin gene expression generates mRNAs encoding different polypeptide products. *Nature* 298:240–244.

Amunts, K., Schlaug, G., Jäncke, L., Steinmetz, H., Schleicher, A., Dabringhaus, A., and Zilles, K. 1997. Motor cortex and hand motor skills: Structural compliance in the human brain. *Human Brain Mapping* 5:206–215.

Anderson, P. T. 2000. *Magnolia: The shooting script.* New York: Newmarket Press.

Anonymous. 1970. Effects of sexual activity on beard growth in man. *Nature* 226:869–870.

Ariew, A. 1996. Innateness and canalization. *Proceedings of the Philosophy of Science Association* 63:19–27.

Aristotle. See Barnes, J.

Baldwin, J. M. 1896. A new factor in evolution. *American Naturalist* 30:441–451, 536–553.

Barker, G. 1993. Models of biological change: Implications of three studies of "Lamarckian" change. In Bateson, P. P. G., Klopfer, P. H., and Thompson, N. S., eds. *Perspectives in ethology.* Vol. 10, *Behavior and evolution.* New York: Plenum Press.

Barnes, J., ed. 1984. *The complete works of Aristotle: The revised Oxford translations.* Vol. 1. Princeton, N.J.: Princeton University Press.

Baxter, L. R., Jr., Schwartz, J. M., Bergman, K. S., Szuba, M. P., Guze, B. H., Mazziotta, J. C., Alazraki, A., Selin, C. E., Ferng, H., Munford, P., and Phelps, M. E. 1992. Caudate glucose metabolic rate changes with both drug and behavior therapy for obsessive-compulsive disorder. *Archives of General Psychiatry* 49:681–689.

Beauchamp, G. K., Doty, R. L., Moulton, D. G., and Mugford, R. A. 1976. The pheromone concept in mammalian communication: A critique. In Doty, R. L., ed. *Mammalian olfaction, reproductive processes, and behavior.* New York: Academic Press.

Beekmans, K., Thiery, E., Derom, C., Vernon, P. A., Vlietinck, R., and Derom, R. 1993. Relating type of placentation to later intellectual development in monozygotic (MZ) twins. *Behavior Genetics* 23:547–548.

Bem, S. L. 1994. *The lenses of gender: Transforming the debate on sexual inequality.* New Haven, Conn.: Yale University Press.

Bickel, H., Gerrard, J., and Hickmans, E. M. 1954. The influence of phenylalanine intake on the chemistry and behavior of a phenylketonuric child. *Acta Paediatrica* 43:64–77.

Billings, P. R., Beckwith, J., and Alper, J. S. 1992. The genetic analysis of human behavior: A new era? *Social Science and Medicine* 35:227–238.

Blakemore, C., and Cooper, G. F. 1970. Development of the brain depends on the visual environment. *Nature* 228:477–478.

Block, N. 1995. How heritability misleads about race. *Cognition* 56:99–128.

Bluestein, H. B. 1999. *Tornado alley: Monster storms of the Great Plains.* New York: Oxford University Press.

Bouchard, T. J., Jr., Lykken, D. T., McGue, M., Segal, N. L., and Tellegen, A. 1990. Sources of human psychological differences: The Minnesota Study of twins reared apart. *Science* 250:223–228.

Bowler, P. J. 1984. *Evolution: The history of an idea.* Los Angeles: University of California Press.

Breedlove, S. M. 1997. Sex on the brain. *Nature* 389:801.

Bruer, J. T. 1999. *The myth of the first three years: A new understanding of early brain development and lifelong learning.* New York: Free Press.

Bull, J. J. 1980. Sex determination in reptiles. *Quarterly Review of Biology* 55:3–21.

Burke, W., Thomson, E., Khoury, M. J., McDonnell, S. M., Press, N., Adams, P. C., Barton, J. C., Beutler, E., Brittenham, G., Buchanan, A., Clayton, E. W., Cogswell, M. E., Meslin, E. M., Motulsky, A. G., Powell, L. W., Sigal, E., Wilfond, B. S., and Collins, F. S. 1998. Hereditary hemochromatosis: Gene discovery and its implications for population-based screening. *Journal of the American Medical Association* 280:172–178.

Carlson, E. A. 1966. *The gene: A critical history.* Philadelphia: W. B. Saunders.

Castle, W. E., and Phillips, J. C. 1909. A successful ovarian transplantation in the guinea-pig, and its bearing on problems of genetics. *Science* 30:312–313.

———. 1913. Further experiments on ovarian transplantation in guinea-pigs. *Science* 38:783–786.

Clark, M. M., Tucker, L., and Galef, B. G. 1992. Stud males and dud males: Intra-uterine position effects on the reproductive success of male gerbils. *Animal Behaviour* 43:215–221.

Clayman, C. B., ed. 1989. *American Medical Association encyclopedia of medicine.* New York: Random House.

Clutter, M. E. 1978. *Dormancy and developmental arrest.* New York: Academic Press.

Cohen, L. G., Celnik, P., Pascual-Leone, A., Corwell, B., Faiz, L., Dambrosia, J., Honda, M., Sadato, N., Gerloff, C., Catala, M. D., and Hallett, M. 1997. Functional relevance of cross-modal plasticity in blind humans. *Nature* 389:180–183.

Cole, M., and Cole, S. R. 1993. *The development of children.* 2d ed. New York: W. H. Freeman.

Collins, F. S. 1999. Shattuck Lecture: Medical and societal consequences of the human genome project. *New England Journal of Medicine* 341:28–37.

Coren, S. 2000. *How to speak dog.* New York: Free Press.

Cossins, A. 1998. Cryptic clues revealed. *Nature* 396:309–310.

Crenshaw, E. B., III, Russo, A. F., Swanson, L. W., and Rosenfeld, M. G. 1987. Neuron-specific alternative RNA processing in transgenic mice expressing a metallothionein-calcitonin fusion gene. *Cell* 49:389–398.

Crutcher, K. A. 1995. The ethics of fetal tissue grafting should be considered along with the science. *Behavioral and Brain Sciences* 18:53–54.

Darwin, C. 1991. *On the origin of species by means of natural selection.* Amherst, N.Y.: Prometheus Books. Originally published in 1859.

Dawkins, R. 1976. *The selfish gene.* Oxford: Oxford University Press.

de Beer, G. R. 1958. *Embryos and ancestors.* 3d ed. London: Oxford University Press.

DeCasper, A. J., and Fifer, W. P. 1980. Of human bonding: Newborns prefer their mothers' voices. *Science* 208:1174–1176.

DeCasper, A. J., and Spence, M. J. 1986. Prenatal maternal speech influences newborns' perception of speech sounds. *Infant Behavior and Development* 9:133–150.

Denenberg, V. H., and Rosenberg, K. M. 1967. Nongenetic transmission of information. *Nature* 216:549–550.

Denenberg, V. H., and Whimbey, A. E. 1963. Behavior of adult rats is modified by experiences their mothers had as infants. *Science* 142:1192–1193.

Dennett, D. C. 1995. *Darwin's dangerous idea: Evolution and the meanings of life.* New York: Simon and Schuster.

Diamond, A., Prevor, M. B., Callender, G., and Druin, D. P. 1997. Prefrontal cortex cognitive deficits in children treated early and continuously for PKU. *Monographs of the Society for Research in Child Development*, serial no. 252, 62 (4).

Dobson, V., and Teller, D. Y. 1978. Visual acuity in human infants: A review and comparison of behavioral and electrophysiological studies. *Vision Research* 18:1469–1483.

Dobzhansky, T. 1955. *Evolution, genetics, and man.* New York: Wiley.

Edelman, G. M. 1992. *Bright air, brilliant fire: On the matter of the mind.* New York: Basic Books.

Elbert, T., Pantev, C., Wienbruch, C., Rockstroh, B., and Taub, E. 1996. Increased cortical representation of the fingers of the left hand in string players. *Science* 270:305–307.

Ellis, L., Ames, M. A., Peckham, W., and Burke, D. 1988. Sexual orientation of human offspring may be altered by severe maternal stress during pregnancy. *Journal of Sex Research* 25:152–157.

Eriksson, P. S., Perfilieva, E., Björk-Eriksson, T., Alborn, A., Nordborg, C., Peterson, D. A., and Gage, F. H. 1998. Neurogenesis in the adult human hippocampus. *Nature Medicine* 4:1313–1317.

Fairchild, H. H. 1991. Scientific racism: The cloak of objectivity. *Journal of Social Issues* 47:101–115.

Fantz, R. L., Ordy, J. M., and Udelf, M. S. 1962. Maturation of pattern vision in infants during the first six months. *Journal of Comparative and Physiological Psychology* 55:907–917.

Ferguson, M. W. J., and Joanen, T. 1982. Temperature of egg incubation determines sex in *Alligator mississippiensis. Nature* 296:850–853.

Flynn, J. R. 1987. Massive IQ gains in 14 nations: What IQ tests really measure. *Psychological Bulletin* 101:171–191.

Garstang, W. 1922. The theory of recapitulation: A critical re-statement of the biogenetic law. *Journal of the Linnean Society of London, Zoology* 35:81–101.

Gearhart, J. 1998. New potential for human embryonic stem cells. *Science* 282:1061–1062.

Gilbert, S. F. 1992a. Cells in search of community: Critiques of Weismannism and selectable units in ontogeny. *Biology and Philosophy* 7:473–487.

———. 1992b. Synthesizing embryology and human genetics: Paradigms regained. *American Journal of Human Genetics* 51:211–215.

———. 1994. *Developmental Biology.* 4th ed. Sunderland, Mass.: Sinauer.

Gilbert, S. F., and Jorgensen, E. M. 1998. Wormholes: A commentary on K. F. Schaffner's "Genes, behavior, and developmental emergentism." *Philosophy of Science* 65:259–266.

Glaser, R., Kennedy, S., Lafuse, W. P., Bonneau, R. H., Speicher, C., Hillhouse, J., and Kiecolt-Glaser, J. K. 1990. Psychological stress-induced modulation of interleukin 2 receptor gene expression and interleukin 2 production in peripheral blood leukocytes. *Archives of General Psychiatry* 47:707–712.

Glaser, R., Rice, J., Sheridan, J., Fertel, R., Stout, J., Speicher, C., Pinsky, D., Kotur, M., Post, A., Beck, M., and Kiecolt-Glaser, J. 1987. Stress-related immune suppression: Health implications. *Brain, Behavior, and Immunity* 1:7–20.

Gleick, J. 1987. *Chaos: Making a new science.* New York: Viking Press.

Gottesman, I. I. 1963. Genetic aspects of intelligent behavior. In Ellis, N., ed. *The handbook of mental deficiency.* New York: McGraw-Hill.

Gottlieb, G. 1965. Imprinting in relation to parental and species identification by avian neonates. *Journal of Comparative and Physiological Psychology* 59:345–356.

———. 1968. Prenatal behavior of birds. *Quarterly Review of Biology* 43:148–174.

———. 1981. Roles of early experience in species-specific perceptual development. In Aslin, R. N., Alberts, J. R., and Petersen, M. R., eds. *Development of perception: Psychobiological perspectives.* Vol. 1, *Audition, somatic perception, and the chemical senses.* New York: Academic Press.

———. 1991a. Experiential canalization of behavioral development: Theory. *Developmental Psychology* 27:4–13.

———. 1991b. Experiential canalization of behavioral development: Results. *Developmental Psychology* 27:35–39.

———. 1992. *Individual development and evolution: The genesis of novel behavior.* New York: Oxford University Press.

———. 1997. *Synthesizing nature-nurture: Prenatal roots of instinctive behavior.* Mahwah, N.J.: Lawrence Erlbaum Associates.

———. In press. Emergence of the developmental manifold concept from an epigenetic analysis of instinctive behavior. In Lewkowicz, D. J., and Lickliter, R., eds. *Conceptions of development: Lessons from the laboratory.* London: Psychology Press.

Gottlieb, G., Wahlsten, D., and Lickliter, R. 1998. The significance of biology for human development: A developmental psychobiological systems view. In Damon, W., and Lerner, R. M., eds. *Handbook of child psychology.* 5th ed. Vol. 1, *Theoretical models of human development.* New York: Wiley.

Gould, E., Beylin, A., Tanapat, P., Reeves, A., and Shors, T. J. 1999. Learning enhances adult neurogenesis in the hippocampal formation. *Nature Neuroscience* 2:260–265.

Gould, S. J. 1977. *Ontogeny and phylogeny*. Cambridge, Mass.: Belknap Press.

———. 1985. *The flamingo's smile: Reflections in natural history*. New York: Norton.

Gray, R. 1992. Death of the gene: Developmental systems strike back. In Griffiths, P., ed. *Trees of life: Essays in philosophy of biology*. Boston: Kluwer Academic Publishers.

Greenough, W. T., Black, J. E., and Wallace, C. S. 1987. Experience and brain development. *Child Development* 58:539–559.

Greenough, W. T., Cohen, N. J., and Juraska, J. M. 1999. New neurons in old brains: Learning to survive? *Nature Neuroscience* 2:203–205.

Griffiths, P. E., and Gray, R. D. 1994. Developmental systems and evolutionary explanation. *Journal of Philosophy* 91:277–304.

Guerra, N. G. 1994. Violence prevention. *Preventive Medicine* 23:661–664.

Hackett, T. 1989. Fire. *New Yorker* 65 (33):50–73.

Haeckel, E. 1866. *Generelle Morphologie der Organismen: Allgemeine Grundzüge der organische Formen-Wissenschaft, mechanisch Begründet durch die von Charles Darwin reformirte Descendenz-Theorie*. 2 vols. Berlin: Georg Reimer.

———. 1874. *Anthropogenie: Keimes- und Stammes-Geschichte des Menschen*. Leipzig: W. Engelmann.

Hall, B. K. 1988. The embryonic development of bone. *American Scientist* 76:174–181.

Hamburger, V. 1934. The effects of wing bud extirpation on the development of the central nervous system in chick embryos. *Journal of Experimental Zoology* 68:449–494.

Herrnstein, R. J., and Murray, C. 1994. *The bell curve: Intelligence and class structure in American life*. New York: Free Press.

Hirsch, H. V. B., and Spinelli, D. N. 1970. Visual experience modifies distribution of horizontally and vertically oriented receptive fields in cats. *Science* 168:869–871.

His, W. 1888. On the principles of animal morphology. *Proceedings of the Royal Society of Edinburgh* 15:287–298.

Ho, M. 1984. Environment and heredity in development and evolution. In Ho, M., and Saunders, P. T., eds. *Beyond neo-Darwinism: An introduction to the new evolutionary paradigm*. London: Academic Press.

Ho, M. W., Tucker, C., Keeley, C., and Saunders, P. T. 1983. Effects of successive generations of ether treatment on penetrance and expression of the *bithorax* phenocopy in *Drosophila melanogaster*. *Journal of Experimental Zoology* 225:357–368.

Hollyday, M., and Hamburger, V. 1976. Reduction of the naturally occurring motor neuron loss by enlargement of the periphery. *Journal of Comparative Neurology* 170:331–320.

Huxley, A. 1939. *After many a summer dies the swan*. New York: Harper.

Immelmann, K. 1969. Song development in the zebra finch and other estrildid finches. In Hinde, R. A., ed. *Bird vocalizations: Their relations to current problems in biology and psychology.* Cambridge, England: Cambridge University Press.

Jablonka, E., and Lamb, M. J. 1995. *Epigenetic inheritance and evolution: The Lamarckian dimension.* Oxford: Oxford University Press.

James, W. 1907. *Pragmatism.* New York: New American Library.

Johannsen, W. 1911. The genotype conception of heredity. *American Naturalist* 45:129–159.

Johnson, M. H. 1997. *Developmental cognitive neuroscience.* Cambridge, Mass.: Blackwell.

Johnston, T. D. 1987. The persistence of dichotomies in the study of behavioral development. *Developmental Review* 7:149–182.

Jones, A. P., and Dayries, M. 1990. Maternal hormone manipulations and the development of obesity in rats. *Physiology and Behavior* 47:1107–1110.

Jones, A. P., and Friedman, M. I. 1982. Obesity and adipocyte abnormalities in offspring of rats undernourished during pregnancy. *Science* 215:1518–1519.

Jones, A. P., and Olster, D. H. 2000. Effects of prenatal, obesity-producing, hormonal and nutritional manipulations on maternal and fetal insulin. Paper presented at the meeting of the Society for the Study of Ingestive Behavior, Dublin, Ireland, July 2000.

Jost, A. 1953. Problems of fetal endocrinology: The gonadal and hypophyseal hormones. *Recent Progress in Hormone Research* 8:379–418.

Kaas, J. H. 1991. Plasticity of sensory and motor maps in adult mammals. *Annual Review of Neuroscience* 14:137–167.

Kagan, J. 2000. The modern synthesis in psychological development. In L. R. Bergman, R. B. Cairns, L. Nilsson, and L. Nystedt, eds. *Developmental science and the holistic approach.* Mahwah, N.J.: Lawrence Erlbaum Associates.

Kagan, J., Kearsley, R. B., and Zelazo, P. R. 1980. *Infancy: Its place in human development.* Cambridge, Mass.: Harvard University Press.

Kandel, E. R., Schwartz, J. H., and Jessel, T. M., eds. 1995. *Essentials of neural science and behavior.* Norwalk, Conn.: Appleton and Lange.

Katz, L. C., and Shatz, C. J. 1996. Synaptic activity and the construction of cortical circuits. *Science* 274:1133–1138.

Keller, E. F. 2000. *The century of the gene.* Cambridge, Mass.: Harvard University Press.

Keller, L., and Ross, K. G. 1993. Phenotypic plasticity and "cultural transmission" of alternative social organization in the fire ant *Solenopsis invicta. Behavioral Ecology and Sociobiology* 33:121–129.

Kelso, J. A. S. 2000. Principles of dynamic pattern formation and change for a science of human behavior. In Bergman, L. R., Cairns, R. B., Nilsson, L., and Nystedt, L., eds. *Developmental science and the holistic approach.* Mahwah, N.J.: Lawrence Erlbaum Associates.

Kempermann, G., Kuhn, H. G., and Gage, F. H. 1997. More hippocampal neurons in adult mice living in an enriched environment. *Nature* 386:493–495.

Kevles, D. J. 1995. *In the name of eugenics.* Cambridge, Mass.: Harvard University Press.

Kleck, R. E., Richardson, S. A., and Ronald, L. 1974. Physical appearance cues and interpersonal attraction in children. *Child Development* 45:305–310.

Kollar, E. J., and Fisher, C. 1980. Tooth induction in chick epithelium: Expression of quiescent genes for enamel synthesis. *Science* 207:993–995.

Kujala, T., Alho, K., Paavilainen, P., Summala, H., and Näätänen, R. 1992. Neural plasticity in processing of sound location by the early blind: An event-related potential study. *Electroencephalography and Clinical Neurophysiology* 84:469–472.

Lamarck, J. B. 1914. *Zoological philosophy: An exposition with regard to the natural history of animals.* Chicago: University of Chicago Press. Originally published in 1809.

Langlois, J. H., Ritter, J. M., Casey, R. J., and Sawin, D. B. 1995. Infant attractiveness predicts maternal behaviors and attitudes. *Developmental Psychology* 31:464–472.

Langman, J., and Wilson, D. B. 1982. Embryology and congenital malformations of the female genital tract. In Blaustein, A., ed. *Pathology of the female genital tract.* 2d ed. New York: Springer-Verlag.

Lehrman, D. S. 1953. A critique of Konrad Lorenz's theory of instinctive behavior. *Quarterly Review of Biology* 28:337–363.

Levine, J., and Suzuki, D. T. 1998. *The secret of life: Redesigning the living world.* New York: W. H. Freeman.

Lewontin, R. C. 1976. Race and intelligence. In Block, N. J., and Dworkin, G., eds. *The IQ Controversy: Critical Readings.* New York: Pantheon.

———. 1983. Gene, organism, and environment. In Bendall, D. S., ed. *Evolution from molecules to men.* Cambridge, England: Cambridge University Press.

———. 1991. *Biology as ideology: The doctrine of DNA.* New York: HarperCollins.

———. 2000. *The triple helix: Gene, organism, and environment.* Cambridge, Mass.: Harvard University Press.

Lewontin, R. C., Rose, S., and Kamin, L. J. 1984. *Not in Our Genes: Biology, Ideology, and Human Nature.* New York: Pantheon.

Lickliter, R., and Berry, T. D. 1990. The phylogeny fallacy: Developmental psychology's misapplication of evolutionary theory. *Developmental Review* 10:348–364.

Lieberman, E. 1995. Low birth weight: Not a black-and-white issue. *New England Journal of Medicine* 332:117–118.

Lindvall, O., Brundin, P., Widner, H., Rehncrona, S., Gustavii, B., Frackowiak, R., Leenders, K. L., Sawle, G., Rothwell, J. C., Marsden, C. D., and Björklund, A. 1990. Grafts of fetal dopamine neurons survive and improve motor function in Parkinson's disease. *Science* 247:574–577.

Maccoby, E. E., and Jacklin, C. N. 1974. *The psychology of sex differences.* Stanford, Calif.: Stanford University Press.

Macfarlane, A. 1978. What a baby knows. *Human Nature* 1:74–81.

Mackie, J. L. 1965. Causes and conditions. *American Philosophical Quarterly* 2:245–264.

Maddox, J. 1991. Is homosexuality hard-wired? *Nature* 353:13.

Magnusson, D. 2000. The individual as the organizing principle in psychological inquiry: A holistic approach. In Bergman, L. R., Cairns, R. B., Nilsson, L., and Nystedt, L., eds. *Developmental science and the holistic approach.* Mahwah, N.J.: Lawrence Erlbaum Associates.

Mann, C. C. 1994. Behavioral genetics in transition. *Science* 264:1686–1689.

Masataka, N. 1993. Effects of experience with live insects on the development of fear of snakes in squirrel monkeys, *Saimiri sciureus. Animal Behaviour* 46:741–746.

Maugh, T. H., II. 2000. Gene therapy may have cured 3 infants. *Los Angeles Times,* April 28:A1.

Maynard Smith, J. 1993. *The theory of evolution.* Canto ed. Cambridge, England: Cambridge University Press.

Mayr, E. 1963. *Animal species and evolution.* Cambridge, Mass.: Harvard University Press.

Meisel, R. L., and Ward, I. L. 1981. Fetal female rats are masculinized by male littermates located caudally in the uterus. *Science* 213:239–242.

Mello, C. V., Vicario, D. S., and Clayton, D. F. 1992. Song presentation induces gene expression in the songbird forebrain. *Proceedings of the National Academy of Sciences of the United States* 89:6818–6822.

Melnick, M., Myrianthopoulos, N. C., and Christian, J. C. 1978. The effects of chorion type on variation in IQ in the NCPP twin population. *American Journal of Human Genetics* 30:425–433.

Merzenich, M. 1998. Long-term change of mind. *Science* 282:1062–1063.

Merzenich, M. M., Nelson, R. J., Stryker, M. P., Cynader, M. S., Schoppmann, A., and Zook, J. M. 1984. Somatosensory cortical map changes following digit amputation in adult monkeys. *Journal of Comparative Neurology* 224: 591–605.

Meulenberg, P. M. M., and Hofman, J. A. 1990. Maternal testosterone and fetal sex. *Journal of Steroid Biochemistry and Molecular Biology* 39:51–54.

Meyer, W. J., Migeon, B. R., and Migeon, C. J. 1975. Locus on human X chromosome for dihydrotestosterone receptor and androgen insensitivity. *Proceedings of the National Academy of Sciences* 72:1468–1472.

Michel, G. F. 1986. Experiential influences on hormonally dependent ring dove parental care. *Annals of the New York Academy of Sciences* 474:158–169.

Michel, G. F., and Moore, C. L. 1995. *Developmental psychobiology: An interdisciplinary science.* Cambridge, Mass.: Massachusetts Institute of Technology Press.

Molnar, Z., and Blakemore, C. 1991. Lack of regional specificity for connections formed between thalamus and cortex in co-culture. *Nature* 351:475.

Moore, C. L. 1992. The role of maternal stimulation in the development of sexual behavior and its neural basis. *Annals of the New York Academy of Sciences* 662:160–177.

Moore, C. L., Dou, H., and Juraska, J. M. 1992. Maternal stimulation affects the number of motor neurons in a sexually dimorphic nucleus of the lumbar spinal cord. *Brain Research* 572:52–56.

Neimark, J. 1997. Nature's clones. *Psychology Today* 30 (4):36–44, 64–69.

Nelkin, D., and Lindee, M. S. 1995. *The DNA mystique: The gene as a cultural icon.* New York: W. H. Freeman.

Neumann-Held, E. M. 1998. The gene is dead—long live the gene: Conceptualizing genes the constructionist way. In Koslowski, P., ed. *Sociobiology and bioeconomics: The theory of evolution in biological and economic theory.* Berlin: Springer-Verlag.

Neville, H. J., Schmidt, A., and Kutas, M. 1983. Altered visual-evoked potentials in congenitally deaf adults. *Brain Research* 266:127–132.

Newman, H. H., Freeman, F. N., and Holzinger, K. J. 1937. *Twins: A study of heredity and environment.* Chicago: University of Chicago Press.

Odling-Smee, F. J. 1988. Niche-constructing phenotypes. In Plotkin, H. C., ed. *The role of behavior in evolution.* Cambridge, Mass.: Massachusetts Institute of Technology Press.

Oyama, S. 1985. *The ontogeny of information: Developmental systems and evolution.* Cambridge, England: Cambridge University Press.

———. 1991. Bodies and minds: Dualism in evolutionary theory. *Journal of Social Issues* 47:27–42.

———. 1992. Transmission and construction: Levels and the problem of heredity. In Tobach, E., and Greenberg, G., eds. *Levels of social behavior: Evolutionary and genetic aspects: Award winning papers from the Third T. C. Schneirla Conference: Evolution of social behavior and integrative levels.* Wichita, Kan.: T. C. Schneirla Research Fund.

Pantev, C., Oostenveld, R., Engelien, A., Ross, B., Roberts, L. E., and Hoke, M. 1998. Increased auditory cortical representation in musicians. *Nature* 392:811–813.

Pascual-Leone, A., and Torres, F. 1993. Plasticity of the sensorimotor cortex representation of the reading finger in Braille readers. *Brain* 116:39–52.

Pedersen, P. E., and Blass, E. M. 1982. Prenatal and postnatal determinants of the first suckling episode in albino rats. *Developmental Psychobiology* 15:349–355.

Penfield, W. 1975. *The mystery of the mind: A critical study of consciousness and the human brain.* Princeton, N.J.: Princeton University Press.

Penfield, W., and Jasper, H. 1954. *Epilepsy and the functional anatomy of the human brain.* Boston: Little, Brown.

Phoenix, C. H. 1974. Prenatal testosterone in the nonhuman primate and its consequences for behavior. In Friedman, R. C., Richart, R. M., and Vande Wiele, R. L., eds. *Sex differences in behavior.* New York: Wiley.

Platt, S. A., and Sanislow, C. A. 1988. Norm of reaction: Definition and misinterpretation of animal research. *Journal of Comparative Psychology* 102:254–261.

Plomin, R. 1994. *Genetics and experience: The interplay between nature and nurture.* Thousand Oaks, Calif.: Sage.

Prigogine, I. 1980. *From being to becoming.* San Francisco: W. H. Freeman.

Querleu, D., and Renard, K. 1981. Les perceptions auditives du foetus humain. *Medicine and Hygiene* 39:2102–2110.

Random House College Dictionary. Rev. ed. 1988. New York: Random House.

Ravelli, G. P., Stein, Z. A., and Susser, M. W. 1976. Obesity in young men after famine exposure in utero and early pregnancy. *New England Journal of Medicine* 295:249–253.

Recio-Pinto, E., and Ishii, D. N. 1988. Insulin and related growth factors: Effects of the nervous system and mechanism for neurite growth and regeneration. *Neurochemistry International* 12:397–414.

Resnick, S. M., Gottesman, I. I., and McGue. M. 1993. Sensation seeking in opposite-sex twins: An effect of prenatal hormones? *Behavior Genetics* 23:323–329.

Ressler, R. H. 1962. Parental handling in two strains of mice reared by foster parents. *Science* 137:129–130.

———. 1963. Genotype-correlated parental influences in two strains of mice. *Journal of Comparative and Physiological Psychology* 56:882–886.

———. 1966. Inherited environmental influences on the operant behavior of mice. *Journal of Comparative and Physiological Psychology* 61:264–267.

Riese, M. L. 1999. Prenatal influences on neonatal temperament: Effects of chorion type for monozygotic twins. Paper presented at the meeting of the Society for Research in Child Development, Albuquerque, N. Mex., April.

Rogoff, B., and Morelli, G. A. 1994. Cross-cultural perspectives on children's development. In Bock, P. K., ed. *Handbook of psychological anthropology.* Westport, Conn.: Greenwood Press.

Rosen, K. M., McCormack, M. A., Villa-Komaroff, L., and Mower, G. D. 1992. Brief visual experience induces immediate early gene expression in the cat visual cortex. *Proceedings of the National Academy of Science of the United States* 89:5437–5441.

Rusak, B., Robertson, H. A., Wisden, W., and Hunt, S. P. 1990. Light pulses that shift rhythms induce gene expression in the suprachiasmatic nucleus. *Science* 248:1237–1240.

Rutherford, S. L., and Lindquist, S. 1998. Hsp90 as a capacitor for morphological evolution. *Nature* 396:336–342.

Sanberg, P. R. 1990. Students' views on fetal neural tissue transplantation. *Lancet* 335:1594.

Scarr, S. 1992. Developmental theories for the 1990s: Development and individual differences. *Child Development* 63:1–19.

Schaffner, K. F. 1998. Genes, behavior, and developmental emergentism: One process, indivisible? *Philosophy of Science* 65:209–252.

Schwartz, J. M., Stoessel, P. W., Baxter, L. R., Jr., Martin, K. M., and Phelps, M. E. 1996. Systematic changes in cerebral glucose metabolic rate after successful behavior modification treatment of obsessive-compulsive disorder. *Archives of General Psychiatry* 53:109–113.

Schwegel, J. 1997. *The baby name countdown.* 4th ed. New York: Marlowe and Company.

Shields, J. 1962. *Monozygotic twins, brought up apart and brought up together.* London: Oxford University Press.

Shorey, M. L. 1909. The effect of the destruction of peripheral areas on the differentiation of the neuroblasts. *Journal of Experimental Zoology* 7:25–63.

Smith, C. W. J., Patton, J. G., and Nadal-Ginard, B. 1989. Alternative splicing in the control of gene expression. *Annual Review of Genetics* 23:527–577.

Smith, L. B. 1999. Do infants possess innate knowledge structures? The con side. *Developmental Science* 2:133–144.

Sokol, D. K., Moore, C. A., Rose, R. J., Williams, C. J., Reed, T., and Christian, J. C. 1995. Intrapair differences in personality and cognitive ability among young monozygotic twins distinguished by chorion type. *Behavior Genetics* 25:457–465.

Solter, D., and Gearhart, J. 1999. Putting stem cells to work. *Science* 283:1468–1470.

Sperry, R. W. 1965. Embryogenesis of behavioral nerve nets. In DeHaan, R. L., and Ursprung, H., eds. *Organogenesis.* New York: Holt, Rinehart, and Winston.

Steele, E. J., Lindley, R. A., and Blanden, R. V. 1998. *Lamarck's signature.* Reading, Mass.: Perseus Books.

Stein, D. A., and Glasier, M. M. 1995. Some practical and theoretical issues concerning fetal brain tissue grafts as therapy for brain dysfunctions. *Behavioral and Brain Sciences* 18:36–45.

Sterelny, K., and Griffiths, P. E. 1999. *Sex and death: An introduction to philosophy of biology.* Chicago: University of Chicago Press.

Stern, K., and McClintock, M. K. 1998. Regulation of ovulation by human pheromones. *Nature* 392:177–179.

Stewart, I. 1989. *Does God play dice? The mathematics of chaos.* Cambridge, Mass.: Basil Blackwell.

Strouse, J. 1980. *Alice James: A biography.* Boston: Houghton-Mifflin.

Sturtevant, A. H. 1915. The behavior of the chromosomes as studied through linkage. *Zeitschrift für Induktive Abstammungs und Verersbungslehre* 13:234–287.

Sur, M. 1993. Cortical specification: Microcircuits, perceptual identity, and overall perspective. *Perspectives on Developmental Neurobiology* 1:109–113.

Sur, M., Pallas, S. L., and Roe, A. W. 1990. Cross-modal plasticity in cortical development: Differentiation and specification of sensory neocortex. *Trends in Neuroscience* 13:227–233.

Susser, E. S., and Lin, S. P. 1992. Schizophrenia after prenatal exposure to the Dutch Hunger Winter of 1944–1945. *Archives of General Psychiatry* 49:983–988.

Tate, D., and Gibson, G. 1980. Socioeconomic status and Black and White intelligence revisited. *Social Behavior and Personality* 8:233–237.

Teicher, M. H., and Blass, E. M. 1976. Suckling in newborn rats: Eliminated by nipple lavage, reinstated by pup saliva. *Science* 193:422–425.

———. 1977. First suckling response of the newborn albino rat: The roles of olfaction and amniotic fluid. *Science* 198:635–636.

Thain, M., and Hickman, M., eds. 1994. *The Penguin Dictionary of Biology.* 9th ed. New York: Penguin Books.

Thelen, E., and Smith, L. B. 1994. *A dynamic systems approach to the development of cognition and action*. Cambridge, Mass.: Massachusetts Institute of Technology Press.

———. 1998. Dynamic systems theories. In Damon, W., and Lerner, R. M., eds. *Handbook of child psychology*. 5th ed. Vol. 1, *Theoretical models of human development*. New York: Wiley.

Thelen, E., and Ulrich, B. D. 1991. Hidden skills: A dynamic systems analysis of treadmill stepping during the first year. *Monographs of the Society for Research in Child Development*, serial no. 223, 56 (1).

Thomson, J. A., Itskovitz-Eldor, J., Shapiro, S. S., Waknitz, M. A., Swiergiel, J. J., Marshall, V. S., and Jones, J. M. 1998. Embryonic stem cell lines derived from human blastocysts. *Science* 282:1145–1147.

Toran-Allerand, C. D. 1976. Sex steroids and the development of the newborn mouse hypothalamus and preoptic area in vitro: Implications for sexual differentiation. *Brain Research* 106:407–412.

Trut, L. N. 1999. Early canid domestication: The farm-fox experiment. *American Scientist* 87:160–169.

Turkewitz, G. 1993. The influence of timing on the nature of cognition. In Turkewitz, G., and Devenny, D. A., eds. *Developmental time and timing*. Hillsdale, N.J.: Lawrence Erlbaum Associates.

van Praag, H., Kempermann, G., and Gage, F. H. 1999. Running increases cell proliferation and neurogenesis in the adult mouse dentate gyrus. *Nature Neuroscience* 2:266–270.

Waddington, C. H. 1957. *The strategy of the genes: A discussion of some aspects of theoretical biology*. New York: Macmillan.

———. 1975. *The evolution of an evolutionist*. Ithaca, N.Y.: Cornell University Press.

Wallman, J. 1979. A minimal visual restriction experiment: Preventing chicks from seeing their feet affects later responses to mealworms. *Developmental Psychobiology* 12:391–397.

Ward, I. L. 1972. Prenatal stress feminizes and demasculinizes the behavior of males. *Science* 175:82–84.

Ward, I. L., and Weisz, J. 1980. Maternal stress alters plasma testosterone in fetal males. *Science* 207:328–329.

Warner, R. R. 1984. Mating behavior and hermaphroditism in coral reef fishes. *American Scientist* 72:128–136.

Watson, J. B. 1930. *Behaviorism*. Rev. ed. Chicago: University of Chicago Press.

Webster's Third New International Dictionary. 1965. Springfield, Mass.: Merriam-Webster.

Weisel, T. N., and Hubel, D. H. 1965. Comparison of the effect of unilateral and bilateral eye closure on cortical unit responses in kittens. *Journal of Neurophysiology* 28:1029–1040.

Weismann, A. 1894. *The effect of external influences upon development*. London: Frowde.

West, M. J., and. King, A. P. 1987. Settling nature and nurture into an ontogenetic niche. *Developmental Psychobiology* 20:549–562.

Wolpert, L. 1991. *The triumph of the embryo.* New York: Oxford University Press.

Wright, L. 1995. Double mystery. *New Yorker* 71 (23): 45–62.

Wright, R. 1995. The biology of violence. *New Yorker* 71 (3): 68–77.

Wright, S. 1920. The relative importance of heredity and environment in determining the piebald pattern of guinea pigs. *Proceedings of the National Academy of Sciences of the United States* 6:320–332.

Yamamoto, K. R. 1985. Steroid receptor regulated transcription of specific genes and gene networks. *Annual Review of Genetics* 19:209–252.

Zigler, E., Taussig, C., and Black, K. 1992. Early childhood intervention: A promising preventative for juvenile delinquency. *American Psychologist* 47:997–1006.

INDEX

Abnormal development, 88–89

Accounting for variation, vs. explaining causation, 41–47

Acquired vs. inherited traits, 41–43, 182–184

Adults
brain plasticity in, 141–143
experience-related effects in, 133–143

African Americans, IQ scores of, 217–221

After Many a Summer Dies the Swan (A. Huxley), 193–194

Aggression, psychobiology of, 226–227

Alper, J. S., 222–223

Alternative splicing, 79

Amara, S. G., 80

Amino acids, 73–74
protein shape and, 75

Amunts, K., 97

Anal-genital licking, in rats, 132, 150

Androgen insensitivity syndrome, 111

Animal cloning, 6–8, 237–241

Aristotle, 16–17, 21

Assembly call of mallard ducklings, 121–122, 188–189, 190

Auditory cortex, sensory processing in, 95–96

Axon(s)
growth cones in, 94
hormonal effects on, 113
synapses and, 97–100

Axon guidance, 93–94

Baer, Karl Ernst von, 18, 19

Baron, David, 239–240

Bases, nucleotide, 72–73

Beard growth, hormones and, 129, 135–136

Beer, Gavin de, 194, 197, 200

Behavior
biological basis of, 13–14

"geneticization of," 222–223

"instinctive," pre/postnatal influences on, 121–122, 130–133

psychobiology of, 222–227

Behavior genetics, 62–63

Behaviorism, 61–62

The Bell Curve (Herrnstein & Murray), 216, 219

Bickel, H., 146

Bidirectionality, in developmental systems, 149

Biological traits, vs. psychological traits, 13–14

Birdsong, 138, 170–171

Birth defects, prenatal factors in, 117–118

Blair, Tony, 208

Blakemore, C., 93–95, 102

Blastula, 84–85

Blending inheritance, 25

Blindness, effects on the brain of congenital, 96

Block, Ned, 45

Bonnie (cloned sheep), 237

Bowler, P. J., 221

Brain
auditory cortex of, 95–96
electrical stimulation of, 91–92
feminization/masculinization of, 112–113
functional organization of, 95–97
hemispheric specialization in, 126–128
plasticity of, 94–96, 141–143
rewiring of, in ferrets, 95
sensory processing in, 95–96
visual cortex of, 95–96, 98–99

Brain development, 91–103
axon guidance in, 93–94
experiential effects in, 94–100
insulin and, 119–121